Biology for Students

Authored by

Mohammad Mehdi Ommati

Henan Key Laboratory of Environmental and Animal Product Safety
College of Animal Science and Technology
Henan University of Science and Technology
Luoyang, Henan 471000, China

Biology for Students

Author: Mohammad Mehdi Ommati

ISBN (Online): 978-981-5324-66-2

ISBN (Print): 978-981-5324-67-9

ISBN (Paperback): 978-981-5324-68-6

First published in 2025.

need for a court order if at any point you breach any terms of this License Agreement. In no event will any delay or failure by Bentham Science Publishers in enforcing your compliance with this License Agreement constitute a waiver of any of its rights.

3. You acknowledge that you have read this License Agreement, and agree to be bound by its terms and conditions. To the extent that any other terms and conditions presented on any website of Bentham Science Publishers conflict with, or are inconsistent with, the terms and conditions set out in this License Agreement, you acknowledge that the terms and conditions set out in this License Agreement shall prevail.

Bentham Science Publishers Pte. Ltd.
80 Robinson Road #02-00
Singapore 068898
Singapore
Email: subscriptions@benthamscience.net

**BENTHAM
SCIENCE**

CONTENTS

FOREWORD I i

FOREWORD II ii

PREFACE AND DEDICATION TO THE FIRST EDITION iii

INTRODUCTION viii

SECTION 1 CELL BIOLOGY - STRUCTURE, FUNCTION, AND DIVISION

CHAPTER 1 EXPLORING THE INNER WORKINGS OF CELLS: UNDERSTANDING THE STRUCTURE AND FUNCTION OF CELLULAR COMPONENTS.................... 1
    UNDERSTANDING THE CHAPTER'S FOCUS: A ROADMAP.......................... 2
    1. CYTOPLASM: A DYNAMIC AND MOBILE CENTER................................. 2
    2. NUCLEUS: THE CELL'S CONTROL CENTER CONTAINING DNA 2
    3. ORGANELLES: DEDICATED CELLULAR UNITS 2
        3.1. Ribosomes and Protein Synthesis 2
        3.2. Endoplasmic Reticulum: Protein and Lipid Synthesis 3
        3.3. Vacuoles and Lysosomes: Cellular Storage and Recycling 3
        3.4. Mitochondria and the Endosymbiotic Theory........................... 4
    4. THE CYTOSKELETON: CELLULAR SUPPORT AND MOVEMENT 4
    5. CELLULAR MOVEMENTS: EXPLORING CELL MOBILITY 5
    COMPREHENSION TASKS 12
    FURTHER DETAILS 20
        Frequently Used Symbols.. 20
    READING UNDERSTANDING 22
    TEXTUAL RESOURCES 28
        Cell Structure and ATP Production 28

CHAPTER 2 CELL DIVISION (CELLULAR REPRODUCTION): EXPLORING THE INTRICACIES OF MITOSIS AND MEIOSIS 31
    UNDERSTANDING THE CHAPTER'S FOCUS: A ROADMAP.......................... 31
    1. THE NUCLEUS AND CHROMOSOMES.. 32
    2. THE CELL CYCLE.. 32
    3. MITOSIS: SEGREGATING THE GENETIC MATERIAL 33
    4. CYTOKINESIS: SEGREGATING THE CYTOPLASM 34
    5. MEIOSIS: FACILITATING SEXUAL REPRODUCTION 35
    6. ASEXUAL *VERSUS* SEXUAL REPRODUCTION 36
    COMPREHENSION TASKS 39
    READING UNDERSTANDING 47
        Reading Material: What Is Biology? 49
        The Importance of Biology.. 50
        Biological Challenges 39

SECTION 2 FOUNDATIONS OF GENETICS - HISTORY AND MOLECULAR BASIS

CHAPTER 3 FROM ANCIENT NOTIONS TO MODERN INSIGHTS: UNRAVELING THE FOUNDATIONS OF GENETICS 52
    UNDERSTANDING THE CHAPTER'S FOCUS: A ROADMAP.......................... 52
    1. EARLY THEORIES OF INHERITANCE... 53
    2. GREGOR MENDEL: PIONEER OF GENETIC INQUIRY 53
    3. MENDEL'S SEMINAL INVESTIGATIONS 53
    4. MENDEL'S INSIGHTS AND THE PRINCIPLE OF INDEPENDENT ASSORTMENT 55
    5. MENDEL'S CONTRIBUTIONS AND REDISCOVERY 55
    6. CHROMOSOMES AND THE FOUNDATIONS OF MENDELIAN GENETICS.............. 56
    COMPREHENSION TASKS 59
    READING UNDERSTANDING 65

Reading Material: Genes and Traits 67

CHAPTER 4 THE CHEMICAL FOUNDATIONS AND STRUCTURAL UNVEILING OF THE GENE 69
    UNDERSTANDING THE CHAPTER'S FOCUS: A ROADMAP................................. 69
    1. EXPLORING THE CHEMICAL BASIS OF GENETIC EXPRESSION 70
    2. PURSUING THE CHEMICAL COMPOSITION AND STRUCTURAL CHARACTERISTICS OF NUCLEIC ACIDS 70
    3. THE RACE FOR UNRAVELING THE MOLECULAR STRUCTURE OF DNA 71
    4. MECHANISMS OF DNA REPLICATION 72
    COMPREHENSION TASKS 77
    READING UNDERSTANDING 84
        Unraveling the Precision of DNA Replication.. 86
            1. Exploring the Mystery of DNA's Accuracy 86
            2. Unveiling the Identity of the Genetic Code 87

SECTION 3 THE DIVERSITY OF LIFE - DEVELOPMENT, ECOLOGY, AND EVOLUTION

CHAPTER 5 LIFE'S STORY: FROM START TO DIVERSITY 89
    UNDERSTANDING THE CHAPTER'S FOCUS: A ROADMAP................................. 89
    1. HOME FOR LIFE: THE BIRTH OF OUR SOLAR SYSTEM AND PLANET EARTH...... 90
    2. THE RISE OF LIFE: FROM CHEMICAL BUILDING BLOCKS TO PROTO-CELLS...... 90
    3. EARLY LIFE FORMS: TRACING THE ORIGIN OF CELLS............................ 91
    4. EARTH'S DYNAMIC EVOLUTION: SHAPING LIFE'S JOURNEY 92
    5. UNVEILING NATURE'S DIVERSITY: THE SCIENCE OF TAXONOMY 92
    6. MAPPING EVOLUTIONARY LINEAGES: THE TRANSITION FROM FIVE KINGDOMS TO THREE DOMAINS 93
    CONCLUSION WITH EXAMPLES IN COW AND MOUSE................................. 94
    COMPREHENSION TASKS.. 101
    READING UNDERSTANDING 107
    READING MATERIALS 109
        I. Earth: The Stage for Life.. 109
            1. How the Earth Formed: from Big Bang to Big Rock 110
            2. Origins of Earth: A Blak and Blue World 110
            3. Earth's Favorable Position in the Solar System.. 111
            4. The Ancient Earth and the Origins of Life's Ingredients.......................... 112
        II. The Invisible Saga: From Molecules to Cells 112
            1. Polymerization: The Formation of Long Molecular Chains 113
            2. Self-Replication: The Emergence of the RNA World 113
            3. The Evolution of Molecular Interactions: From RNA to Genetic Information Transmission 114
            4. Emergence of Cell-like Structures: The Protocell Hypothesis 114
            5. The Emergence of Coordinated Cellular Activities 115

CHAPTER 6 FUNGI: THE MIGHTY DECOMPOSERS 116
    UNDERSTANDING THE CHAPTER'S FOCUS: A ROADMAP................................. 116
    1. FUNGI'S ATTRIBUTES 117
    2. FUNGI CLASSIFICATION 117
    3. LICHENS: NATURE'S PERFECT SYMBIOSIS 119
    4. FUNGAL EVOLUTIONARY INSIGHTS 119
    COMPREHENSION TASKS 122
    READING UNDERSTANDING 129
    READING MATERIALS 132
        Exploring Microbiology: An Overview 132
            1. The Discovery of Microorganisms 133
            2. The Controversy Surrounding Spontaneous Generation 133
            3. The Contribution of Microorganisms to Illness 135
            4. Microbial Applications in Industry and Ecosystem Dynamics 138
            5. Exploring Microbial Diversity: Prokaryotes, Eukaryotes, and Modern Classification 139
            6. Microbiology: Its Scope and Importance 140

7. The Future Of Microbiology 142
CHAPTER 7 THE JOURNEY OF ANIMAL DEVELOPMENT: FROM GAMETE PRODUCTION TO FERTILIZATION 144
    UNDERSTANDING THE CHAPTER'S FOCUS: A ROADMAP 144
    1. GENERATING SPERM AND EGGS 145
    2. FERTILIZATION: THE GENESIS OF DEVELOPMENT 146
    3. CLEAVAGE: THE FOUNDATION OF CELL PROLIFERATION 146
      Summary of Cleavage in Early Development 147
      Species-Specific Differences in Embryonic Development 148
        1. Fertilization and Early Cleavage 148
        2. Morula and Blastocyst Formation 150
        3. Species-Specific Timing of Implantation 150
        4. Evolutionary and Developmental Implications 151
    4. DEVELOPMENTAL MILESTONES: FROM GASTRULATION TO ORGANOGENESIS 151
        4.1. Gastrulation: Orchestrating Embryonic Layers 151
        4.2. Organogenesis: Sculpting Functional Tissues and Organs 152
    5. PROTECTIVE SHELLS: EXTRAEMBRYONIC MEMBRANES IN LAND VERTEBRATES 153
    6. GROWTH DYNAMICS: CELLULAR PROLIFERATION AND DEVELOPMENTAL PHASES 153
    7. AGING DYNAMICS: THE CULMINATION OF DEVELOPMENT 154
    COMPREHENSION TASKS 160
    READING UNDERSTANDING 167
      Diverse Feeding Strategies Among Warbler Species in Spruce-Fir Forests 167
    READING MATERIALS 170
      Animals 170
      1. Exploring the Complexity of Animal Morphology and Embryonic Development 170
      2. Unveiling the Enigmatic World of Sponges: Primitive Creatures Defying Traditional Animal Classification 170
      3. Diversity of the Arthropods 171
        3.1. Chelicerates: Exploring Ancient Lineages 172
        3.2. Mandibulates: Exploring the World of Crustaceans 172
        3.3. Mandibulates: Unraveling the World of Insects 173
      4. Chordates: Unveiling the Essence of Vertebrate Evolution 174
        4.1. Tunicates: Delving into the Primitive Ancestry of Chordates 175
        4.2. Lancelets: Exploring the Fishlike Marvels of Cephalochordates 175
        4.3. Vertebrates: Embarking on the Evolutionary Journey of Craniate Chordates 175
      4.1. Tunicates: Delving into the Primitive Ancestry of Chordates 175

CHAPTER 8 EVOLUTIONARY FORCES SHAPING SPECIES DIVERSITY 177
    UNDERSTANDING THE CHAPTER'S FOCUS: A ROADMAP 177
    1. HOW BIOLOGISTS DEFINE A SPECIES 178
    2. MECHANISMS OF GENE EXCHANGE PREVENTION 178
    3. EVOLUTION INTO SPECIES: ESTABLISHING GENETIC ISOLATION 179
    4. GENETIC FOUNDATIONS OF SPECIATION 179
    5. UNDERSTANDING MACROEVOLUTION: MAJOR TRANSFORMATIONS 179
    6. MICROEVOLUTION'S CONTRIBUTION TO MACROEVOLUTION 181
    COMPREHENSION TASKS 182
    READING UNDERSTANDING 190
      Reading Materials: How Species Form 195
        1. The Nature of Species 196
        2. The Divergence of Populations 197
        3. Reproductive Isolating Mechanisms 198

CHAPTER 9 POPULATION DYNAMICS AND ECOLOGY 200
    UNDERSTANDING THE CHAPTER'S FOCUS: A ROADMAP 200
    1. FACTORS INFLUENCING POPULATION GROWTH 201
    2. CONSTRAINTS ON POPULATION EXPANSION 201
    3. PATTERNS OF POPULATION DISTRIBUTION 202

4. HUMAN POPULATION GROWTH: AN EXPONENTIAL CASE STUDY 205
COMPREHENSION TASKS 207
READING UNDERSTANDING 214
READING MATERIALS 219
   Insights into Behavioral Ecology 219
      1. Understanding Behavior 219
      2. Instinct 219
      3. Learned Behavior 219
      4. Conditioning 220
      5. Imprinting 220
      6. Insight Learning 220

SECTION 4 ADVANCES IN BIOLOGICAL RESEARCH AND INFORMATION SYSTEMS

CHAPTER 10 NAVIGATING CANCER SYSTEMS BIOLOGY 221
  UNDERSTANDING THE CHAPTER'S FOCUS: A ROADMAP 221
  1. CANCER SYSTEMS BIOLOGY AND ITS IMPACT ON PRECISION MEDICINE 222
    1.1. The Transformative Impact of Systems Biology on Cancer Research 222
    1.2. Systems Biology as a Tool for Customized Medicine 223
  2. APPROACHES TO INVESTIGATING CANCER SYSTEMS BIOLOGY 225
    2.1. Essential Models for Cancer Systems Biology 227
    2.2. Integrating Data for the Development of Cancer Gene Networks 228
      2.2.1. Development of Cancer Gene Networks 228
      2.2.2. Advancements in Bioinformatics: Unveiling New Frontiers in Systems Biology Data Integration and Analysis 230
    2.3. Unlocking Complexity: Advancements and Challenges in Network Visualization for Understanding Cancer Molecular Networks 231
    2.4. Navigating the Intricacies of Biological Systems: A Comprehensive Examination of Network Analysis in Cancer Molecular Networks 232
      2.4.1. Different Types of Networks Encode Various Biological Properties 232
      2.4.2. Exploring Biological Systems through Network Analysis 233
      2.4.3. Analyzing Network Dynamics for Diagnosis and Prognosis through Gene Markers 234
    2.5. Dynamic Network Modeling: Deciphering the Information Processing Machinery of Cancer Cells 235
  COMPREHENSION TASKS 238
  READING UNDERSTANDING 247
  Diagnosis of Diseases in Livestock 247
      I. Importance of Diagnosis 247
      II. Diagnostic Approach 247
      III. Post-Diagnosis Actions 248
      IV. Diagnosis of Deceased Animals 248
      V. Preventing Disease Outbreaks 248

CHAPTER 11 THE EVOLUTION AND FUTURE PROSPECTS OF THE ISI WEB OF KNOWLEDGE PLATFORM 252
  UNDERSTANDING THE CHAPTER'S FOCUS: A ROADMAP 252
  1. OVERVIEW 252
  2. THE ROLE OF CITED REFERENCES IN SHAPING SCIENTOMETRICS 253
    2.1. How 'Citation Indexes' Transformed into the 'ISI Web of Science' 254
    2.2. The ISI Web of Knowledge: A Comprehensive Research Tool 255
      2.2.1. Introduction to the ISI Web of Knowledge 255
      2.2.2. Enhancing Research Through Seamless Access 256
      2.2.3. Comprehensive Content and Advanced Tools 256
  3. THE VALUE OF MULTIDISCIPLINARY CONTENT IN ISI WEB OF KNOWLEDGE ... 256
    3.1. Unified Access to Diverse Information 256
    3.2. The Evolution of Global Research Accessibility 257
    3.3. Challenges of Information Overload 257
    3.4. Streamlined Access to Relevant Information 257
    3.5. Evaluated Web Content for Scholarly Use 257

3.6. Expanding Content Offerings through Strategic Partnerships 258
4. ISI WEB OF KNOWLEDGE: LEVERAGING INNOVATIVE TOOLS THROUGH ADVANCED TECHNOLOGIES 258
4.1. Enhancing Research Capabilities with Advanced Tools.. 258
4.2. ISI CrossSearch and eSearch: Unifying Access to Diverse Research Resources.......... 258
4.3. Future Directions for ISI Web of Knowledge... 259
5. ISI LINKS: AN INTEGRATED LINKING SOLUTION 260
5.1. Comprehensive Linking Capabilities.. 260
5.2. Accessing Full-Text: Direct Publisher Connections 261
5.3. ISI RoboLinks: Ensuring No Dead Links 261
5.4. Customization and Appropriate Copy... 262
5.5. SFX for Context-Sensitive Linking: Extending Library Resources.............................. 262
CONCLUSION 262
COMPREHENSION TASKS 265
READING UNDERSTANDING 270
A. Immunology.. 270
B. Immune System 273
C. Cytokines 275
D. Extending Scientific Knowledge 276

SECTION 5 CRAFTING THE SCHOLAR'S PATH - NAVIGATING THE ACADEMIC MANUSCRIPT JOURNEY

CRAFTING THE SCHOLAR'S PATH: NAVIGATING THE ACADEMIC MANUSCRIPT JOURNEY 281

SECTION 6 CONCLUSION

CHAPTER 12 CONCLUSION: THE SYMPHONY OF LIFE – INTEGRATING KNOWLEDGE AND ADVANCING FRONTIERS 343
The Road Ahead: Emerging Fields and Ethical Considerations.................................... 344
Final Thoughts 345

SECTION 7 ANSWERS
ANSWERS.. 346

SECTION 8 APPENDIX
APPENDIX: VOCABULARY PREFIXES AND SUFFIXES ROOTS 382

BIBLIOGRAPHY 419

SUBJECT INDEX 424

FOREWORD I

Dr. Mohammad Mehdi Ommati obtained his M.Sc. and Ph.D. degrees under my supervision in the Department of Animal Science, College of Agriculture, Shiraz University, Shiraz, Iran. I was fortunate to guide such a talented and highly enthusiastic student in his desire to be trained in reproductive physiology. He was a top student amongst his peers throughout his graduate and undergraduate studies, simultaneously being chosen as a member of the National Elites Foundation (INEF), Iran. After successfully completing his sabbatical studies at Shanxi Agricultural University (SXAU), where he served as an assistant professor for about five years, Dr. Ommati joined the College of Animal Science and Technology at Henan University of Science and Technology, China, where he is currently an associate professor. Believing in quality research, Dr. Ommati has always carried out his research with utmost diligence and meticulously. This is borne by the fact that his research findings have been published in many reputable periodicals, and books, and presented at international scientific meetings, as well as receiving several international awards. He is currently on the editorial boards of several reputable journals and is also a visiting professor at the Shiraz University of Medical Sciences, Shiraz, Iran. His recent endeavor is the current book aimed at presenting the basic knowledge and current advancements in biological sciences, making it a useful source for students of general biology, animal science, medicine, and veterinary science. I believe he has all the essential qualities for writing this book. I wish him the best in pursuing his future scientific activities.

M. J. Zamiri
Department of Animal Science,
Shiraz University,
Shiraz, Iran

FOREWORD II

After obtaining his degree in Animal Sciences in 2009, Professor M. Mehdi Ommati initiated a successful career in reproductive physiology. He is a highly qualified researcher, teacher, editorial member, and reviewer in several journals, with very important research in reproductive toxicology. His articles have a high citation index, and he has received awards and recognition in countries such as China, India, Mexico, *etc*. Professor Ommati is an outstanding example of a young researcher who, with great effort, dedication, and commitment, has managed to position himself as one of the most outstanding researchers at an international level, with recognized prestige. His achievements reflect his great capabilities, making him a great motivation for new generation of young researchers. Despite his achievements, Professor Ommati continues to work with great dedication and enthusiasm, inspiring his undergraduate and graduate students to pursue distinguished careers in the biological sciences. In addition, Professor Ommati has always shown the best willingness and ability to establish international collaborations with his academic peers, with a great vision for research. This book is intended for postgraduate students, but it can also be useful for students in biology, animal sciences, and related medical disciplines as a support in their training and to reinforce their knowledge. The lecture of this book is enjoyed while reading and reinforces the basic knowledge that every biological sciences student should know, regardless of their specialization. The glossaries are very complete, the self-assessments reinforce what was learned in each lesson, and the reading comprehension is excellent. It is a very didactic book, an enormous effort to summarize the most important issues about molecules, cells, systems, organisms, and their habitat. My recognition to Professor Ommati for his contribution, not only to the field of reproductive toxicology but also as a generous and grateful teacher for the gifts he has received from God.

Socorro Retana-Márquez
Autonomous Metropolitan University
Mexico City, Mexico

PREFACE AND DEDICATION TO THE FIRST EDITION

The twenty-first century is an era of rapid advancements in biological sciences, marked by continuous breakthroughs in knowledge and innovative methodologies. To support the effective learning and application of this specialized knowledge, and in response to the increasing demand for creativity and innovation in higher education, I have designed this book. It is my hope that it will meet the needs of students and researchers around the world, fostering scientific exchange and collaboration across borders. Drawing inspiration from a range of international references, this book is presented to you, the global community of learners, educators, and researchers.

Though I am physically distant from my homeland, *Iran*, my thoughts and efforts remain firmly anchored in its academic community. Despite being far from home, I have always kept a close eye on the developments within my country, eager to support its students and educators. Even from this distance, I continue to dedicate myself to advancing knowledge through my publications—over 250 articles in prestigious global journals, several books, and chapters in international volumes. It is my sincere hope that these contributions, however modest they may seem, can support and inspire the next generation of scholars and students, both in Iran and beyond.

Exploring the Intricacies of Biology: From Cells to Systems and Beyond

In this book, efforts have been made to cover various topics such as modern biology, the use of new resources, and attention to new developments in the field of biology. As mentioned above, this book addresses aspects of writing research papers, submitting articles, searching for resources, and using scientific sources, somewhat tailored to the level of graduate students. This book can serve as a primary educational resource in universities for students in fields such as biology, animal sciences, veterinary medicine, and other related medical disciplines. It has also been designed to be a reference source for individuals working in fields like biology, agriculture, veterinary medicine, and other related areas.

This book is meticulously structured to appeal to students, educators, and biology enthusiasts. It provides a thorough exploration of fundamental and advanced concepts in four main sections. Each section includes several chapters, and each chapter comprises various chapters. At the end of the book, answers to exercises and a glossary are provided to support your learning further. This comprehensive text covers a variety of topics, including cell biology, genetics, animal biology, microbiology, biochemistry, biosystematics, principles and models of evolution, information science, and writing research papers.

Section One: Cell Biology - Structure, Function, and Division. In this section, I embark on a journey into the microscopic world of cells. **Chapter 1** provides an in-depth understanding of cellular components and their functions, while **Chapter Two** explores the complex processes of mitosis and meiosis, shedding light on cellular reproduction.

Section Two: Foundations of Genetics - History and Molecular Basis. This section traces the evolution of genetic science. **Chapter Three** takes you from ancient theories to modern genetic insights, and **Chapter Four** unravels the chemical and structural foundations of genes, offering a molecular perspective on genetic inheritance.

Section Three: The Diversity of Life - Development, Ecology, and Evolution. Here, I explore the vast diversity of life forms. **Chapter Five** narrates the story of life's evolution, **Chapter Six** examines the crucial role of fungi, and **Chapter Seven** guides you through animal development from gamete production to fertilization. **Chapter Eight** and **Chapter Nine** discuss the evolutionary forces and ecological dynamics shaping species diversity and population ecology.

Section Four: Advances in Biological Research and Information Systems. This section highlights modern advancements in biology. **Chapter Ten** focuses on cancer systems biology, providing insights into the complexities of cancer research, and **Chapter Eleven** explores the evolution and prospects of the ISI Web of Knowledge platform, a crucial tool for modern biological research.

Section Five: Navigating the Academic Manuscript Journey. This section is an invaluable resource for aspiring scholars. **Part A** offers a detailed guide to creating effective tables and figures, discusses the art of crafting a compelling discussion section, and provides tips for developing an effective manuscript draft and selecting the right journal. **Part B** outlines guidelines for manuscript submission, includes a submission checklist, and offers strategies for promoting your publication. **Parts C and D** delve into language editing services, while **Parts E to H** cover comprehensive referencing guides and provide additional information on major research databases like Elsevier ScienceDirect, ISI Web of Knowledge, and Springer Link.

Section Six: Answers. This section provides answers and explanations for all chapters from sections One to Eleven, serving as a crucial resource for self-assessment and deeper understanding.

Section Seven: Appendix. The appendix offers additional resources and references. It breaks down the fundamental components of medical English, focusing on roots, prefixes, and suffixes.

Main Features of this Book:

1. **Expanded Vocabulary and Specialized Terminology:** Critical biological terms and concepts are highlighted in **bold** throughout the chapters, providing immediate recognition for students. These terms are further clarified in the *"Vocabulary (Index of Terms)"* section, located just before the exercises and reading questions, ensuring that students can easily look up definitions while engaging with the content. This approach makes it easier for students to grasp new vocabulary in context.

2. **Development of Core Scientific Skills**: This book not only focuses on delivering biological content but also prioritizes the development of essential academic skills. Throughout the chapters, students are encouraged to read specialized articles, search for scientific information, and write scientific papers. These practices are integrated into the book to prepare students for advanced studies or professional endeavors in the biological sciences.

3. **Practical and Engaging Content**: The chapters cover a broad range of topics that are both practical and interesting, ensuring that the material is both informative and engaging. Students will encounter complex concepts presented in an accessible way, fostering a deeper understanding and sparking curiosity.

4. **Broad and Relevant Coverage**: The content spans diverse and contemporary areas of biology, from molecular biology to ecology, ensuring that students gain a holistic understanding of the discipline. Each chapter is designed to provide insights into the latest trends and advancements, empowering students to stay informed and adaptable in their future careers.

5. **The Brainstorm (Roadmap)**: The "Brainstorm" feature is a key component of each chapter, placed right after the keywords. It serves as a concise summary and guide to the core concepts, helping students gain a clearer understanding of the chapter's objectives. By offering a structured outline of

the chapter's key points, this feature encourages active reading and enhances comprehension, allowing students to better focus on the most important aspects of the material. This roadmap is a unique feature that empowers students to approach complex topics with more confidence and clarity.

By integrating insights from the latest international textbooks, scientific journals, and online resources, this book ensures that students are exposed to the most current knowledge in the field. This educational approach broadens students' understanding of biology, while also fostering their self-learning capabilities, preparing them for success both in their studies and in their professional lives. I hope this book serves as an invaluable resource, helping students build a strong foundation in biology and equipping them for lifelong learning and scientific inquiry.

With deep respect and gratitude, I extend my appreciation to those who will enhance the value of this book with their feedback and suggestions. The sources used in this book are listed in the references (see the last pages: Sources and Additional Resources), and I am obliged to thank the authors of these works. Despite my utmost efforts to present a complete and accurate work, errors and omissions are inevitable. I kindly ask readers to share their opinions and criticisms with me *via* email (mehdi_ommati@outlook.com).

This book was finalized on May 2, 2024, coinciding with my 38th birthday, in Luoyang, China. With the grace of the Almighty, this work has come to an end. I sincerely thank my wife, *Samira Sabouri*, who has always been a guide and a loyal friend with her wisdom and calm demeanor. I am deeply grateful to my son, *Adrian Ommati*, whose thirst for knowledge and big dreams have always been a source of inspiration for me, and to my newborn daughter, *Ariana Ommati*, whose pure and beautiful presence brings me peace and joy. May their presence always inspire me to create remarkable works. Finally, I dedicate this book to my dear parents, *Hossein Ommati* and *Akram Piroozfar*, and wish them health from afar.

The Path of Love and Sacrifice

As I the close this preface, I would like to share with you, dear readers, a couplet by the great Persian poet Hafez that beautifully captures the depth of love and sacrifice:

"The path of love is such that it has no borders, There is no solution but to give up one's life in it."

Hafez, a revered poet of Iran, speaks of the "path of love" in this couplet—a path without borders or an end, where the only way to reach the destination is through self-sacrifice and dedication. This concept holds true not only in literature but also in life and science. The love for knowledge and research is also an endless journey that requires relentless effort and sacrifice. This couplet, like a guiding light, reminds us that in the pursuit of knowledge and progress, we must never shy away from dedication and hard work. It is my hope that this book can serve as a companion on this path filled with love and sacrifice, aiding you every step of the way.

Mohammad Mehdi Ommati

Henan Key Laboratory of Environmental and Animal Product Safety
College of Animal Science and Technology
Henan University of Science and Technology
Luoyang, Henan 471000, China

INTRODUCTION

Chapter Overview

Welcome to a comprehensive journey through the multifaceted world of biological sciences, where foundational principles meet cutting-edge research and technological advancements. This book is structured to guide readers through a broad spectrum of topics, from the molecular underpinnings of life to the intricate dynamics of ecosystems, and from evolutionary processes to the latest in cancer systems biology and bibliometrics. Each chapter stands as a testament to the complexity and beauty of life, offering insights that are both profound and practical.

In Chapter one, we delve into the core principles that form the bedrock of biological understanding. "Chapter One: Cell Biology - Structure, Function, and Division" begins with an exploration of the inner workings of cells. Here, we uncover the structure and function of cellular components, laying a foundation that is essential for understanding more complex biological processes. We then move on to the intricacies of cellular reproduction, examining the processes of mitosis and meiosis, which are fundamental to the continuity of life.

"*Chapter Two*: Foundations of Genetics - History and Molecular Basis" takes us through the historical evolution of genetic thought, from ancient notions to modern insights. We explore the chemical foundations and structural complexities of genes, providing a thorough understanding of how genetic information is encoded, transmitted, and expressed. This chapter sets the stage for comprehending the profound impact of genetics on all living organisms.

The third chapter, "The Diversity of Life - Development, Ecology, and Evolution," celebrates the vast array of life forms and their developmental processes. We trace life's story from its origins to the diverse forms we see today, including a detailed look at fungi, the mighty decomposers of the natural world. This chapter also covers the journey of animal development, from gamete production to fertilization, and the evolutionary forces that shape species diversity. Understanding population dynamics and ecology, we learn how species interact with their environments and each other, maintaining the delicate balance of ecosystems.

In "*Chapter Four*: Advances in Biological Research and Information Systems," we transition to the forefront of biological research. We explore the realm of cancer systems biology, where a systems-level approach is essential to understanding and treating this multifacetcd disease. The evolution and future prospects of the ISI Web of Knowledge platform are discussed, highlighting its role in revolutionizing how we access and analyze scientific literature.

Section Five of the book, "Crafting the Scholar's Path - Navigating the Academic Manuscript Journey," serves as a practical guide for aspiring researchers. It offers step-by-step instructions for crafting effective Tables and Figures, developing a robust first draft of your manuscript, and selecting the appropriate journal for publication. Detailed guidelines for writing and submitting manuscripts, a journal submission checklist, and strategies for promoting your publication are also provided. Additionally, this section includes insights into language editing services and comprehensive guides on referencing, including the Harvard Referencing System and additional resources such as Elsevier ScienceDirect and Springer Link.

Section Seven, "*Answers*," provides comprehensive answers to the questions and tasks presented throughout the lessons, reinforcing key concepts and ensuring a deep understanding of the material.

Finally, *Section Eight, "Appendix*," includes valuable supplementary materials. The appendices feature a general vocabulary list and an exploration of the fundamental components of medical English vocabulary, including roots, prefixes, and suffixes. These resources are designed to support readers in mastering the specialized language of biology and medicine.

One core innovation of this book is the **Brainstorm (Roadmap)** feature, placed after each chapter's keywords. This section offers a structured guide that highlights key ideas, connections to other topics, and the broader context, helping students navigate complex material and enhancing comprehension.

Book Objectives

This book aims to provide a thorough and accessible exploration of contemporary biological sciences. By bridging foundational knowledge with the latest research and technological developments, it offers readers a comprehensive understanding of the field. Whether you are a student, educator, researcher, or enthusiast, this book will serve as a valuable resource for deepening your knowledge and appreciation of the living world.

Conclusion

As we embark on this journey through the chapters, we invite you to delve into the intricacies of life, from the molecular to the ecological level. Each chapter not only imparts essential knowledge but also inspires curiosity and a deeper understanding of the natural world with vocabulary and reading sections that help the readers to extend their knowledge and skills. The interconnectedness of life is a central theme, highlighting the complexity and beauty of biological systems and the ever-evolving nature of scientific discovery.

Section 1

Cell Biology - Structure, Function, and Division

CHAPTER 1

Exploring the Inner Workings of Cells: Understanding the Structure and Function of Cellular Components

Abstract: This chapter delves into the complex and intricate structures within cells, focusing on the cytoplasm, nucleus, organelles, cytoskeleton, and cellular movements. It begins by examining the dynamic nature of the cytoplasm, highlighting its role as the mobile center of cellular activity. The nucleus is discussed as the central hub of genetic information, enclosed by the nuclear envelope. The chapter then explores various organelles, including plastids, ribosomes, lysosomes, vacuoles, mitochondria, the endoplasmic reticulum, and the Golgi complex, each contributing uniquely to cellular functions. The cytoskeleton's role in providing support and facilitating movement within the cell is analyzed, followed by an exploration of cellular mobility mechanisms, including chemotaxis, cilia, and flagella. Plant cell-specific processes like cytoplasmic streaming are also covered, providing a comprehensive overview of cellular components and their functions.

Keywords: Cytoplasm, Cytoskeleton, Nucleus, Organelles, Protein synthesis.

UNDERSTANDING THE CHAPTER'S FOCUS: A ROADMAP

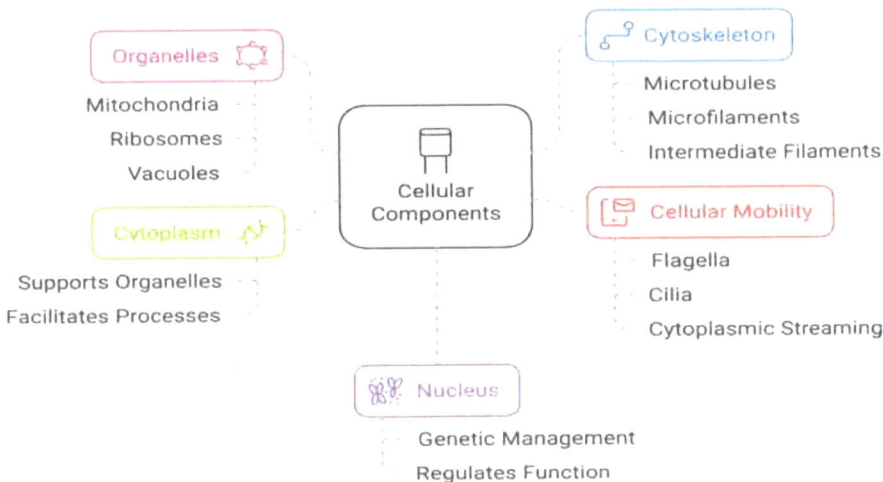

1. CYTOPLASM: A DYNAMIC AND MOBILE CENTER

The cytoplasm exhibits numerous vital characteristics of life, functioning as a dynamic and mobile center within cells. This semifluid substance constitutes a significant proportion of cellular mass and is enclosed by the plasma membrane. Within the cytoplasm, organelles are suspended and supported structurally by the filamentous **cytoskeleton**. Additionally, the cytoplasmic fluid contains vital nutrients, soluble proteins, ions, and other key substances necessary for proper cellular function.

2. NUCLEUS: THE CELL'S CONTROL CENTER CONTAINING DNA

The **nucleus**, a prominent organelle in eukaryotic cells, serves as the central hub for genetic information. As the largest organelle, it stores the cell's genetic material, DNA, which is organized into chromosomes (in prokaryotic cells, genetic material is located in the nucleoid). The nucleus also contains one or more nucleoli, which play a critical role in ribosome biogenesis and are essential for cell division. The nucleus is enclosed by a porous double membrane called the **nuclear envelope**, which separates it from the cytoplasm. While small molecules can diffuse freely across this membrane, larger molecules like mRNA and ribosomal subunits require specialized nuclear pores for transport.

3. ORGANELLES: DEDICATED CELLULAR UNITS

Specialized units known as organelles are present in all eukaryotic cells, each performing specific roles essential for cellular activity. Organelles such as mitochondria, the Golgi apparatus, the endoplasmic reticulum, vacuoles, ribosomes, lysosomes, and plastids (in plant cells) are responsible for diverse functions (Fig. **1**).

3.1. Ribosomes and Protein Synthesis

Ribosomes, critical for protein production, can number from a few hundred to several thousand within a single cell. Composed of a large and a small subunit, ribosomes link amino acids to form proteins during translation, moving along mRNA strands. Groups of ribosomes can attach to a single mRNA strand, forming **polysomes**. Proteins synthesized in the cytoplasm are used within the cell, whereas those intended for export or membrane integration are produced in conjunction with the rough endoplasmic reticulum (RER).

3.2. Endoplasmic Reticulum: Protein and Lipid Synthesis

The **endoplasmic reticulum (ER)** is a complex system of interconnected sacs, tubules, and vesicles that occurs in two forms: rough (RER) and smooth (SER). The RER, studded with ribosomes, specializes in protein synthesis and transport, while the SER focuses on lipid synthesis and detoxification of harmful substances. Additionally, the RER is involved in the formation of the nuclear envelope following cell division.

Transport vesicles carry molecules from the ER to the **Golgi apparatus**, where they are further modified and packaged for export or intracellular distribution.

3.3. Vacuoles and Lysosomes: Cellular Storage and Recycling

Vacuoles, which appear as empty sacs under a microscope, are filled with fluid and dissolved substances. In plant cells, large central vacuoles store water, nutrients, and waste products, while in animal cells, vacuoles are involved in processes such as **phagocytosis** (engulfing particles) and **pinocytosis** (cellular intake of liquids).

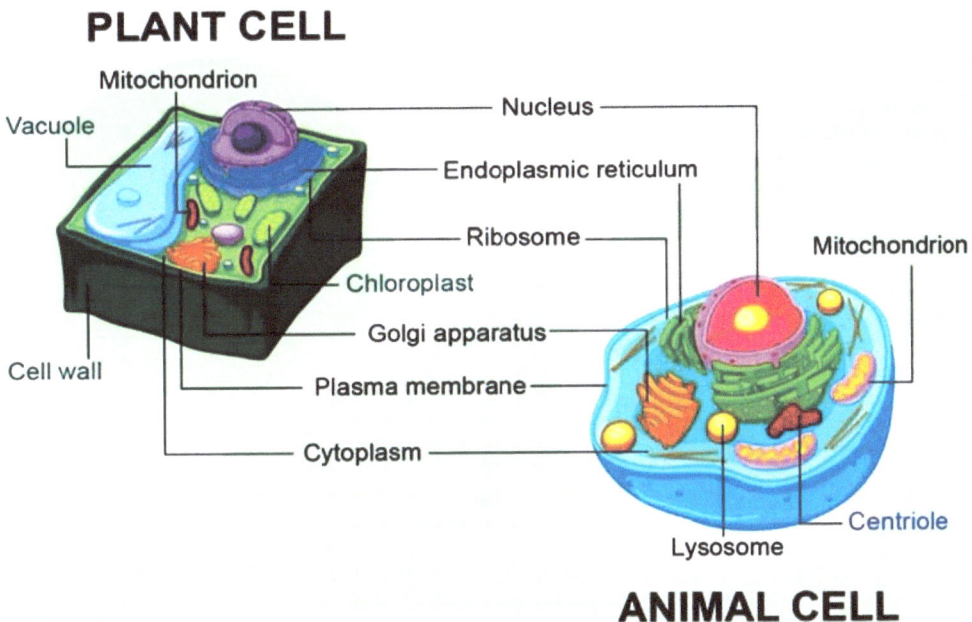

Fig. (1). Structural and functional differences between plant and animal cells.

(A) Plant Cell:

•Enclosed by a cellulose-based cell wall for structural support.

•Contains a large central vacuole and chloroplasts for photosynthesis.

•Typically has a regular, rectangular shape due to the cell wall.

(B) Animal Cell:

•Lacks a cell wall, enabling flexibility.

•Contains smaller vacuoles and no chloroplasts.

•Often round or irregular in shape.

Lysosomes, a specialized type of vacuole, contain digestive enzymes synthesized by the RER and processed by the Golgi apparatus. These enzymes break down macromolecules, process food particles, and recycle damaged cellular components.

3.4. Mitochondria and the Endosymbiotic Theory

Mitochondria, known as the cell's powerhouses, produce ATP *via* enzymes located on the folds of their inner membranes (cristae). These organelles are self-replicating, supporting the widely accepted "Endosymbiotic Theory", which proposes that mitochondria originated as free-living prokaryotes that formed a symbiotic relationship with early eukaryotic cells.

In plant cells, **chloroplasts**—a type of **plastid**—facilitate photosynthesis, converting light energy into chemical energy. Chloroplasts contain stacks of membranes called grana within a fluid matrix known as the **stroma**.

4. THE CYTOSKELETON: CELLULAR SUPPORT AND MOVEMENT

The cytoskeleton, a dynamic network of filaments and tubules, provides structural support to eukaryotic cells (Fig. **2**). Comprising mainly threadlike **microfilaments** composed of **actin**, the cytoskeleton facilitates movement within cells across both plant and animal types. Additionally, the contractile protein **myosin** is involved in the contraction of muscle cells.

Microtubules, which are another essential part of the cytoskeleton, are made of **tubulin** protein and serve as a structural framework that helps maintain cell shape. Intermediate filaments enhance the overall strength and stability of the cytoplasm.

Dynein, myosin, and kinesin (known as mechanoenzymes) engage with elements of the cytoskeleton to produce forces that drive cellular motion.

5. CELLULAR MOVEMENTS: EXPLORING CELL MOBILITY

While the cytoskeleton contributes to cellular stability, its microtubules, filaments, and associated proteins enable various types of locomotion, such as crawling or sliding. These types of movement generally need a firm surface for the cells to attach to and can be influenced by the surface's shape. Furthermore, certain cells demonstrate **chemotaxis**, which is the capacity to move toward or away from chemical gradients.

Some eukaryotic cells can swim freely in liquid environments, driven by whip-like appendages known as **cilia** or **flagella**. Both structures possess a similar internal arrangement, featuring nine pairs of microtubules organized in a circular formation, along with two central microtubules. Movement is facilitated by dynein side arms extending from microtubules within each doublet, anchored at the cell surface by **basal bodies**.

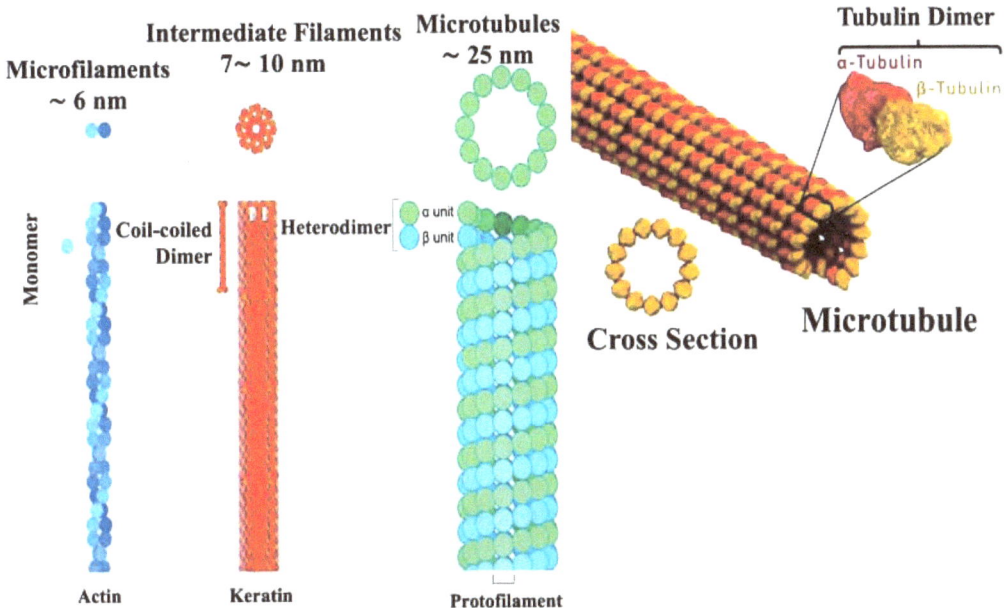

Fig. (2). Molecular components of the cytoskeleton.

In plant cells, nutrients, proteins, and other materials are transported *via* cytoplasmic streaming. Myosin proteins connected to organelles exert force on microfilaments spread throughout the cell, aiding in movement. Both microfilaments and microtubules are essential for nearly all significant cytoplasmic activities. In cell division, spindle microtubules, formed from tubulin subunits located near the centrioles, are responsible for transporting chromosomes to support the division process.

Vocabulary (Index of Terms)

Actin ['æktɪn]

Actin is a globular contractile protein found in cells, particularly abundant in muscle cells. It is a component of the cytoskeleton and plays a role in several cellular functions, such as cell movement, cell division, and the preservation of cell shape. While actin is most notably associated with muscle contraction, it is also present in non-muscle cells, where it plays roles in cell motility, cytoplasmic streaming, and intracellular transport.

Basal body ['beɪsəl 'bɒdi]

A structure similar to a centriole that is consistently found at the bottom of a cilium or a eukaryotic flagellum. It serves as an anchor point for the cilium or flagellum, facilitating cellular movement and sensory functions.

Centriole ['sɛntriˌoʊl]

A cylindrical organelle is usually located close to the nucleus in many animal cells and some lower plant cells, but it is not present in prokaryotic cells or higher plants. Centrioles play a crucial role in organizing the microtubules during cell division, contributing to the formation of the mitotic spindle and ensuring proper chromosome segregation.

Chemotaxis[ˌkiːməˈtæksɪs]

The purposeful movement of an organism or cell in reaction to an external chemical signal. This movement is typically aimed at finding or avoiding specific substances in the environment.

Chloroplast ['klɔːrəˌplæst]

A type of plastid responsible for conducting photosynthesis. Chloroplasts are found in all organisms that perform photosynthesis, excluding certain types of photosynthetic bacteria and cyanobacteria. They contain chlorophyll and other pigments that capture light energy and convert it into chemical energy through photosynthesis.

Chromosome ['kroʊmə͵soʊm]

A filamentous structure made up of DNA and histone proteins, located inside the nucleus. Generally, each chromosome features two telomeres at either end, a centromere located centrally, and may also include a nucleolus organizer region. Chromosomes play a critical role in storing and transmitting genetic information from one generation to the next. They are associated with RNA and various proteins essential for gene expression and regulation.

Cilia ['sɪliə]

Hair-like structures protrude from the cell surface, often in large numbers, facilitating cellular movement. Cilia play essential roles in locomotion, such as propelling fluids over cell surfaces or enabling the movement of single cells through their environment.

Cytoplasm ['saɪtə͵plæzəm]

The gel-like substance fills the interior of a cell, surrounded by the cell membrane (plasmalemma). It consists of a watery ground substance called hyaloplasm or cell sap, containing various organelles and particles. The cytoplasm excludes the nucleus and any visible vacuoles. It plays a vital role in cellular metabolism, providing a medium for chemical reactions and serving as a scaffold for organelle positioning.

Cytoskeleton [͵saɪtəʊ'skɛlɪtən]

In eukaryotic cells, the cytoskeleton serves as an internal "skeleton", providing structural support and organization. Composed of microtubules and other components, it maintains cell shape, facilitates intracellular transport, and coordinates cell movement. Additionally, the cytoskeleton enables free-living cells to move through their environment.

Dynein ['daɪni:n]

Dynein describes a family of no fewer than four unique proteins located within the flagella and microtubules of eukaryotic cells. These proteins have the ability to hydrolyze ATP, which enables them to utilize energy from ATP molecules to power cellular movement, particularly along microtubules.

Endoplasmic reticulum [ɛndə'plæzmɪk rɪ'tɪkjʊləm]

The endoplasmic reticulum (ER) consists of folded membranes and tubules spread throughout the eukaryotic cell. It creates an extensive surface area that facilitates a variety of chemical processes, such as synthesizing proteins, metabolizing lipids, and storing calcium. ER is classified into two types: rough ER, which has ribosomes on its surface and is crucial for protein production, and smooth ER, which does not have ribosomes and is involved in lipid production and detoxification functions.

Flagella [flə'dʒɛlə]

Flagella are long, whip-like structures protruding from the cell surface that facilitate cellular movement. They enable locomotion by propelling the cell through fluid environments, such as water or mucus. Flagella are typically fewer in number compared to cilia and are involved in movements such as swimming or directing the movement of fluid over the cell surface.

Golgi complex ['gɒldʒi 'kɒmplɛks]

The Golgi complex is composed of a series of flattened, smooth membrane-bound sacs found in eukaryotic cells. It functions as the location for synthesizing, modifying, and packaging specific molecules, especially proteins and lipids, prior to their transport to their ultimate locations within or outside the cell.

Lysosome ['laɪsə‚soʊm]

Lysosomes are specialized organelles containing a variety of hydrolytic enzymes. These enzymes are involved in breaking down cellular waste materials, such as damaged organelles or macromolecules, as well as engulfed foreign particles, through a process known as hydrolysis. This helps maintain cellular homeostasis and plays a crucial role in cellular digestion and recycling.

Microfilament ['maɪkroʊ‚fɪləmənt]

Microfilaments are elongated, thread-like structures made of protein that are frequently present in cells, often alongside microtubules. They are essential for offering structural support to the cell and aiding in various cellular movements. Microfilaments play a role in several cellular activities, including the maintenance of cell shape, cell division, cell movement, and the transport of materials within the cell.

Microtubule ['maɪkroʊ‚tubjuːl]

Microtubules are tiny, cylindrical structures made up of protein subunits that are spread throughout the cytoplasm of cells. They are vital for offering structural support to the cell and enabling numerous cellular functions, such as cell division, transport of materials within the cell, and cell movement. Microtubules act as pathways along which motor proteins transport organelles and other cellular elements, aiding in the cell's organization and dynamic processes.

Mitochondrion [maɪtoʊˈkɒndrɪən]; Mitochondria [maita'kondria]

Mitochondria are membrane-bound organelles that resemble small sacs, with an inner membrane that folds back on itself. They function as the primary site for aerobic cellular respiration, generating energy through oxidative phosphorylation. Mitochondria play a crucial role in producing adenosine triphosphate (ATP), which serves as the main energy source for cellular activities, by breaking down nutrients in the presence of oxygen. Moreover, mitochondria participate in various other cellular functions, including apoptosis (the process of programmed cell death) and the regulation of calcium signaling.

Myosin ['maɪoʊsɪn]

Myosin is a protein that, in conjunction with actin, constitutes the main element of the contractile machinery within muscle cells. It is crucial for muscle contraction, as it interacts with actin filaments to produce force and facilitate movement. Myosin molecules undergo a series of conformational changes during muscle contraction, enabling them to pull actin filaments closer together and produce muscle contraction.

Nuclear envelope ['njuːklɪər ˈɛnvəˌloʊp]

The nuclear envelope is a structure composed of two lipid bilayers and associated proteins, forming a double membrane that encloses the nucleus of a cell. It serves as the outermost boundary of the nucleus, separating its contents from the cytoplasm. The nuclear envelope controls the transport of substances, such as RNA and proteins, between the nucleus and the cytoplasm nuclear pores embedded within its structure. Additionally, it provides structural support and helps maintain the integrity of the nucleus.

Nucleoid ['njuːklɪɔɪd]

The nucleoid is the DNA-containing region found within prokaryotic cells, serving a similar function to the nucleus in eukaryotic cells but lacking a membrane boundary. It is a condensed region where the genetic material of the cell is localized, allowing for essential cellular processes such as replication, transcription, and translation to occur. The nucleoid is not enclosed by a membrane, distinguishing it from the nucleus of eukaryotic cells.

Nucleoli [njuːˈkliːəˌlaɪ]

Nucleoli are nuclear components found within the cell nucleus, consisting of fully or partially assembled ribosomes and particular chromosomal regions that contain the genetic instructions necessary for their formation. They play a crucial role in ribosome biogenesis, where ribosomal RNA (rRNA) is transcribed, processed, and assembled with ribosomal proteins to form functional ribosomes. Nucleoli are dynamic and undergo changes in response to cellular conditions, such as changes in the rate of protein synthesis or cellular stress.

Nucleus ['njuːklɪəs]; nuclei ['njuːklɪˌaɪ])

The nucleus, an organelle in eukaryotic cells, contains chromosomes that control cellular processes and transmit genetic information, ensuring inheritance *via* DNA. The nucleus functions as the cell's control center, overseeing gene expression, DNA replication, and various critical cellular activities. The nucleus, surrounded by a double membrane called the nuclear envelope, houses nucleoplasm, chromatin, and one or more nucleoli. It performs a vital role in preserving cellular homeostasis and passing genetic information to daughter cells during cell division.

Phagocytosis [fægoʊsaɪˈtoʊsɪs]

Phagocytosis is the cellular process in which a cell surrounds and engulfs a particle, such as a bacterium or foreign material, by forming a membrane-bound vesicle called a phagosome. This process is typically carried out by specialized cells called phagocytes, including macrophages and neutrophils, as part of the immune response to pathogens and foreign substances. Phagocytosis has an important task in defense against infections and in the clearance of cellular debris, contributing to the body's overall immune function.

Pinocytosis [paɪnoʊsaɪˈtoʊsɪs]

Pinocytosis is a cellular process in which a cell engulfs fluid and dissolved molecules from its surroundings by forming small, membrane-bound vesicles called pinocytic vesicles or endosomes. Unlike phagocytosis, which involves the ingestion of large particles or microorganisms, pinocytosis primarily facilitates the uptake of extracellular fluid and solutes, such as nutrients, ions, and signaling molecules. This process occurs in most eukaryotic cells and plays an important role in nutrient uptake, cell volume regulation, and cellular signaling.

Plastid ['plæstɪd]

A plastid is a small, membrane-bound organelle found in almost all plant cells, but absent in fungi, blue-green algae, and bacteria. Enclosed by a double membrane known as the envelope, plastids perform various functions, including photosynthesis, storage, and pigment synthesis.

Polysome ['pɑlɪsoʊm]	In the process of protein synthesis, a polysome refers to several ribosomes working in tandem, translating the same messenger RNA molecule consecutively.
Ribosome ['raɪbə‚soʊm]	Ribosomes are tiny structures made up of two subunits, consisting of proteins and ribonucleic acid (RNA). They play a crucial role in assembling proteins from amino acids during protein synthesis.
Stroma ['stroʊmə]; stromata ['stroʊmətə])	The stroma refers to the region within a chloroplast that lacks chlorophyll pigment.
Tubulin ['tuːbjʊlɪn]	Tubulin is a spherical protein that assembles into elongated, cylindrical structures called microtubules through polymerization. These microtubules are essential parts of the cytoskeleton in eukaryotic cells, offering structural stability, aiding in intracellular transport, and playing significant roles in processes such as cell division, movement, and overall cellular organization. Tubulin exists in several isoforms, allowing for diverse functions and regulation within the cell.
Vacuole ['væKjuːoʊl]	A vacuole is a storage compartment found within the cytoplasm of a cell, enclosed by a membrane. It serves to store various substances such as water, ions, nutrients, and waste products, contributing to cellular homeostasis, turgor pressure regulation, and intracellular digestion. Vacuoles can vary in size, composition, and function across different cell types and organisms.
	In addition to its storage function, the vacuole plays a critical role in autophagy, a cellular mechanism responsible for breaking down and recycling various cellular components. During autophagy, injured organelles, proteins, and other cytoplasmic constituents are sequestered into double-membraned vesicles called autophagosomes. These autophagosomes then fuse with vacuoles, forming autolysosomes where the engulfed material is broken down by hydrolytic enzymes. This process allows the cell to maintain nutrient availability, remove damaged components, and respond to various stresses, contributing to cellular health and survival.

COMPREHENSION TASKS

I. Key Terms: Matching

Instructions: Pair each term in the left column with its appropriate description in the right column. Write the letter of the correct description next to the term's number. Each description will only match with one term. Good luck!

1. Polysome	A. Protein Synthesis
2. Pinocytosis	B. Baglike Structure
3. Exocytosis	C. Power Generator
4. Plastid	D. Where Flagella Grow
5. Golgi Complex	E. Toward or Away From A Chemical Stimulus
6. Flagella	F. Engulfment
7. Phagocytosis	G. RNA and Ribosomes
8. Lysosome	H. Weblike
9. Basal Body	I. In Plants Only
10. Chemotactic	J. Control Room
11. Nucleus	K. Expel
12. Vacuole	L. Vacant
13. Ribosome	M. Whiplike
14. Cytoskeleton	N. Cell Drinking
15. Mitochondrion	O. Packaging

II. Cell Biology True or False Quiz: Determine whether the following statements are true or false based on your understanding of cell biology concepts. Write "True" if the statement is correct and "False" if the statement is incorrect. Good luck!

1.   ----- The DNA of prokaryotic cells is not contained within a nucleus.

2.   ----- Ribosomes are not derived from the nucleoli.

3.   ----- The nuclear envelope has pores, unlike other cell membranes.

4.   ----- The cytoskeleton helps maintain the position of the smooth endoplasmic reticulum.

5.   ----- Structural proteins can be exported from the cell.

6.   ----- The rough endoplasmic reticulum does not generate the nuclear envelope.

7.   ----- The majority of cellular proteins are synthesized on ribosomes.

8.   ----- White blood cells make use of phagocytosis as a process.

9.   ----- Some prokaryotic cells possess microbodies.

10.   ----- Mitochondria have the ability to replicate independently.

11.   ----- Pinocytosis entails the uptake of fluid into a cell through a vacuole.

12.   ----- Prokaryotic and eukaryotic cells have a cytoskeleton for support.

13.   ----- Carotenoids are usually pigmented molecules.

14.   ----- Grana are not enclosed by stomata.

III. Cell Biology Completion Exercise: Fill in the blanks with the appropriate term or phrase to complete each sentence related to cell biology. Write your answers in the space provided. Good luck!

1. Phagocytosis is a process of cellular feeding that first requires the _____ of the food.

2. The lysosome packages some fifty hydrolytic enzymes in ----------.

3. ---------- are lysosome-like vesicles containing waste products. They are thought to be involved with cell ----------.

4. Both ---------- and ---------- are thought to have arisen from endosymbiosis.

5. The cytoskeleton is composed of very fine ----------, medium ---------- and larger -----.

6. Creeping and gliding cell movements are usually ---------- -dependent.

7. ---------- behavior is shown when a cell moves toward or away from a chemical substance.

8. Flagella emerge from the surface of the cell exclusively at the _____.

IV. Multiple Choice: Choose the best answer for each question by circling the corresponding letter.

1. The majority of traits linked to life processes relate to the properties of ____.

a. Endosymbionts

b. The cytoplasm

c. DNA

d. The nucleus

e. None of the above

2. Ribosomes ____.

a. Serve as organelles involved in protein synthesis

b. Act as the cell's main energy source

c. Function as storage sites for starch

d. Participate in the breakdown of proteins

e. Store genetic information in the form of DNA

3. Smooth endoplasmic reticulum (SER) __.

a. Is devoid of ribosomes

b. Is engaged in the synthesis of fats and steroids

c. Plays a role in the detoxification of toxins

d. All of the above

e. None of the above

4.   Ribosomes are synthesized in ___.

a. The cytoplasm

b. Nucleoli

c. Mitochondria

d. Smooth endoplasmic reticulum

e. Rough endoplasmic reticulum

5.   Lysosomes contain _____.

a. Hydrolytic enzymes

b. Genetic material

c. Stored fats

d. Proteins

e. Carbohydrates

6.   The process of phagocytosis entails___.

a. Engulfing solid particles within vacuoles

b. Exocytosis

c. Uptake of water through a cell's vacuole

d. Removal of solid particles from a cell

e. Elimination of water from a cell

7. The main function of ____ is to convert and store energy within the cell.

a. Ribosomes: protein synthesis structures

b. Microbodies: peroxisomes

c. Contractile vacuoles: vacuoles that manage cell pressure

d. Mitochondria: powerhouse organelles

e. Smooth endoplasmic reticulum: non-ribosomal endoplasmic reticulum

8. Enzymes involved in ATP production in mitochondria are primarily located ____.

a. Inside the mitochondrial matrix

b. On the cristae

c. On ribosomes located in the matrix

d. Spread throughout the cristae and matrix

e. Associated with polysomes

9. Chromoplasts are a form of ____.

a. Color pigment

b. Storage structure

c. Nutrient reservoir

d. Plastid

e. None of the options above

10. An mRNA molecule along with its attached ribosomes forms _____.

a. A multisome: a ribosomal cluster

b. A polysome: a ribonucleoprotein complex

c. A lysosome: a digestive organelle

d. A monosome: a single ribosome complex

e. None of the options listed

11. Leucoplasts are a type of plastid that ____.

a. Hold carotenoid pigments

b. Involve in the process of photosynthesis

c. Serve as reservoirs for proteins, fats, and starch

d. is involved in providing color to plants

e. None of the aforementioned options

12. Every cell has a framework of filaments and tubules called the____.

a. Cytosol

b. A vacuole: A storage organelle

c. An endoplasmic reticulum

d. A cytoskeleton structure

e. A cell membrane

13. In prokaryotic cells, the DNA exists____.

a. Located within a nucleus

b. Arranged into distinct chromosomes

c. Compacted into a region known as the nucleoid

d. Grouped within nucleoli

e. Surrounded by a nuclear membrane

14. Proteins are formed from amino acids ____.

a. Inside the nucleus

b. At the sites of ribosomes

c. Within the mitochondria

d. Inside lysosomes

e. In the Golgi apparatus

15. In protein production, one mRNA strand can link to multiple ribosomes, forming ____.

a. A segment of DNA

b. A cellular organelle for digestion

c. A polysome

d. A type of endoplasmic reticulum without ribosomes

e. A type of endoplasmic reticulum with ribosomes

16. Proteins meant for secretion or membrane incorporation can be recognized by their____.

a.   Their secondary structural configuration

b. An amino acid sequence referred to as a signal peptide

c. Their connection to a polysome

d. All of these factors

e. None of these factors

17. Proteins produced on the endoplasmic reticulum undergo modifications ___.

a. Within vacuoles

b. At the cell membrane

c. In the Golgi apparatus

d. Inside lysosomes

e. They are not modified after synthesis

18. An amoeba residing in a freshwater environment, which has a lower tonicity than its internal cytoplasm, must manage a continuous intake of water. It eliminates this surplus water by ___.

a. Employing excretory proteins

b. Sealing its cell membrane with lipids

c. Utilizing a contractile vacuole

d. Engaging in phagocytosis

e. Relocating to an area with lower tonicity

19. The cellular hydrolytic enzymes break down food within the ___.

a. Lysosomal compartment

b. Golgi apparatus

c. Mitochondria: powerhouse of the cell

d. Chloroplast: photosynthetic organelle

e. Membrane network involved in protein and lipid synthesis

20. An organelle obtained from an animal cell is found to possess a significant amount of enzymes that are critical for energy conversion processes. This organelle is probably ___.

a. A lysosome body

b. A mitochondrion

c. A Golgi complex

d. A leucoplast

e. A chloroplast

FURTHER DETAILS

Frequently Used Symbols

1. Chemical Elements

Aluminum	**Al**	Helium	**He**	Potassium	**K**
Argon	**Ar**	Hydrogen	**H**	Radon	**Rn**
Arsenic	**As**	Iron	**Fe**	Molybdenum	**Mo**
Beryllium	**Be**	Krypton	**Kr**	Selenium	**Se**
Boron	**B**	Lead	**Pb**	Silicon	**Si**
Cadmium	**Cd**	Magnesium	**Mg**	Silver	**Ag**

Calcium	**Ca**	Manganese	**Mn**	Sodium	**Na**
Carbon	**C**	Mercury	**Hg**	Sulfur	**S**
Chlorine	**Cl**	Molybdenum	**Mo**	Xenon	**Xe**
Chromium	**Cr**	Neon	**Ne**	Zinc	**Zn**
Cobalt	**Co**	Nickel	**Ni**	-	-
Copper	**Cu**	Nitrogen	**N**	-	-
Fluorine	**F**	Oxygen	**O**	-	-
Gold	**Au**	Phosphorus	**P**	-	-

2. Mathematical Symbols

a squared	a^2	Point (or decimal) zero three	0.03
b cubed	b^3	Four point two	4.2
c to the power of four	c^4	x to the power of five	x^5
f to the power of minus one	f^{-1} or f^(-1)	y squared	y^2
Zero (or nought) point six	0.6	z to the power of six	z^6

3. Symbols commonly employed for measurements: length, capacity, and weight.

Acre [#]	**a**	Litre	**L**
Centimeter	**cm**	Metre	**m**
Cubic centimeter	**cm³**	Micron	**µm**
Cubic decimeter	**dm³**	Milligram	**mg**
Cubic meter	**m³**	Milligramme	**mg**

(Table) cont.....

Cubic millimeter	mm³	Milliliter	mL
Decimeter	dm	Millimeter	mm
Gram	g	Millimetre	mm
Gramme	g	Square centimeter	cm²
Hectare	ha	Square decimeter	dm²
Kilogram	kg	Square kilometer	km²
Kilogramme	kg	Square meter	m²
Kilometer	km	Square millimeter	mm²
Liter	L	Tonne	t

READING UNDERSTANDING

Lesson One: Understanding Bird Migration

Section 1:

Bird migration, or the seasonal movement of birds from one region to another, is a fascinating phenomenon that scientists are still trying to fully grasp. One puzzling aspect is what prompts birds to embark on these journeys. Some experts suggest that birds migrate to escape harsh winter conditions. Their fluffy feathers provide excellent insulation against the cold, and their warm-blooded nature ensures that their body temperature remains constant, regardless of external temperatures.

This instinctive behavior is observed even in young birds. For instance, arctic terns, born in the frigid Arctic, join the migration without ever having seen the distant lands they're flying to. Scientists believe that hormonal changes triggered by the lengthening or shortening of daylight play a crucial role in regulating bird migration. In an experiment, birds exposed to artificial light showed increased gland activity, leading to the secretion of hormones that control various bodily functions.

Interestingly, birds seemed more energetic as the artificial night extended, which aligns with the observation that many migratory flights occur during the night.

Additionally, shorter daylight periods prompt birds to store more fat, serving as fuel for their long journeys. This sheds light on the complex interplay of factors influencing bird migration.

1. What is the primary subject addressed in the text?

A. Common migratory paths for birds

B. Reasons behind bird migration

C. Species of non-migratory birds

D. The movement of birds in frigid regions.

2. What, as stated in the text, helps shield birds from cold temperatures?

A. Glands: secretory organs

B. Hormones: chemical messengers

C. Feathers

D. Man-made illumination

3. In the opening paragraph, what is the nearest definition of the term "constant"?

A. Ongoing

B. Unchanging

C. Anticipatable

D. Dependable

4. In the second paragraph, why are young arctic terns mentioned?

A. They do not migrate

B. They migrate instinctively

C. They have been studied

D. They adapt to cold climates

5. The passage discusses all of the following changes experienced by birds exposed to longer periods of darkness EXCEPT __.

A. Stimulated secretory organs

B. Excited behavior

C. Accumulation of additional fat

D. Enhanced desire to eat

6. What factor did the researchers modify for the birds in the experiment described in the passage?

A. Availability of food

B. Core temperatures

C. Amount of light exposure

D. Neurochemical balance

7. What is the significance of feathers for birds during migration?

A. They provide insulation against cold weather

B. They regulate hormonal changes

C. They serve as fuel for long flights

D. They control body temperature fluctuations

8. According to the passage, what role do hormones play in bird migration?

A. They trigger excitement in birds

B. They regulate gland activity

C. They stimulate appetite

D. They influence flight patterns

9. In the passage, what is suggested as a possible trigger for bird migration?

A. Changes in food availability

B. Hormonal imbalances

C. Fluctuations in daylight length

D. Variations in temperature

10. Based on the information provided, why do scientists believe bird migration is instinctive?

A. Birds travel long distances without stopping

B. Birds migrate even when conditions are favorable

C. Birds show increased gland activity during migration

D. Birds are observed migrating from birth without prior experience

Section 2:

Bees, much like ants, are social creatures that live together in hives. Within these hives, there are three distinct types of bees: the queen, drones, and workers. Each bee plays a vital role in supporting the group. The queen bee, a female, assumes the pivotal task of egg-laying, ensuring the hive's continuity. Contrarily, drones, the male bees, focus on mating with the queen to fertilize her eggs. Additionally, they contribute to the hive by engaging in activities such as foraging for food. Meanwhile, the workers, also female, engage in a multitude of duties essential for hive maintenance and survival. These include gathering nectar (With the help of enzymes, this nectar is transformed into honey), nurturing the queen and larvae, and upholding the hive's cleanliness and structure. Beyond gathering nectar, worker bees also undertake tasks like regulating hive temperature. Each bee's contribution is integral to the hive's thriving ecosystem, exemplifying the remarkable synergy found in bee colonies.

As mentioned, the honey is then stored in the hive's cells, serving as a crucial food source during the winter months. Some individuals are involved in beekeeping, managing numerous hives to collect honey. Once collected, honey is often bottled or jarred for sale. Its color varies depending on the flowers from which the bees source the nectar.

1. Why are bees referred to as social insects? Because they __ .

A. Live collectively

B. Inhabit areas close to humans

C. Require beekeepers

D. Exhibit a diligent work ethic

2. How do bees utilize nectar?

A. Transform it into sugar

B. Convert it into honey

C. Offer it to the queen bee

D. Utilize it in constructing their hives

3. In what containers is honey typically retailed?

A. Beehives

B. Packs of golden color

C. Honey sacks

D. Bottles or jars

4. What determines the color of honey in its final form?

A. The type of flower from which the nectar was gathered

B. The quantity of sugar provided by beekeepers

C. The amount of water accessible to the bees

D. The season during which the nectar was collected

5. What is the primary responsibility of the queen bee in the hive?

A. Collecting nectar

B. Constructing the hive

C. Laying eggs

D. Protecting the hive

6. In addition to mating, what do drones contribute to the hive?

A. Pollinating flowers

B. Nurturing the queen bee

C. Building hive cells

D. Foraging for food

7. What tasks do worker bees perform besides gathering nectar?

A. Regulating hive temperature

B. Guarding the hive entrance

C. Fanning to circulate air

D. Dancing to communicate

8. Why is honey stored in the hive's cells during winter?

A. To attract more bees

B. To defend against predators

C. As a food source

D. To create space for eggs

9. What is the primary role of individuals engaged in beekeeping?

A. Protecting wild bee populations

B. Maintaining bee-friendly habitats

C. Collecting honey from hives

D. Breeding new bee species

10. According to the reading, what influences the color variation of honey?

A. The type of wax used in hives

B. The age of the worker bees

C. The temperature inside the hive

D. The variety of flowers visited by bees

TEXTUAL RESOURCES

Cell Structure and ATP Production

Biological systems, despite lacking the theoretical foundations of physics, display remarkable unity, notably in cellular construction.

1. Water-Impermeable Boundaries:

Cells necessitate borders to confine their contents and preserve biological functions. These boundaries, akin to shells, prevent the escape of cellular contents into the surrounding milieu. Constructed from fatty molecules, cellular membranes act as impermeable barriers, crucial for cell integrity. The arrangement of these molecules forms a double-layered membrane with a fatty interior and water-compatible exterior, akin to oil and water's distinct properties. Cholesterol molecules interspersed within add rigidity to the membrane. While fat-based membranes lack strength, organisms have evolved mechanisms like cell walls in plants and external proteins in mammals to bolster cellular borders.

2. Cellular Diversity:

Two main cell types exist: prokaryotic, lacking a nucleus, and eukaryotic, possessing one. Prokaryotic cells, exemplified by bacteria, have no internal compartments, while eukaryotic cells, including yeast and multicellular organisms, feature nuclei and organelles like lysosomes, each serving distinct functions. The nucleus segregates DNA from the cytoplasm, enabling complex gene regulation, a hallmark of eukaryotic cells' evolutionary advancement.

3. Membrane Functions Beyond Segregation:

Cell membranes not only segregate cellular compartments but also facilitate energy storage and molecular anchoring. By restricting molecule diffusion, membranes store energy analogous to batteries. Moreover, membrane-anchored proteins, like transporters and receptors, regulate molecular transport and intercellular signaling, coordinating cellular activities.

4. Organelle Evolution:

Mitochondria and chloroplasts, organelles vital for energy production, originated from bacteria engulfed by ancestral eukaryotic cells. Despite dependence on nuclear functions, these organelles retain their DNA and bacterial features, highlighting their evolutionary origins and the unified energy mechanisms across all cells.

5. ATP Production Mechanisms:

All living organisms rely on adenosine triphosphate (ATP) as their energy currency. The synthesis of ATP occurs through two main mechanisms:

• Substrate-Level Phosphorylation:

The production of ATP from adenosine diphosphate (ADP) and inorganic phosphate (Pi) requires an input of free energy, rendering the process endergonic and non-spontaneous. However, coupling this synthesis with an exergonic reaction, which releases a surplus of energy, facilitates ATP production. This coupling of reactions, where energy from an exergonic process drives ATP synthesis, is termed substrate-level phosphorylation.

• Chemiosmotic Generation of ATP:

Organisms possess transmembrane channels that expel protons from cells. Excited electrons induce shape changes in membrane proteins, leading to proton extrusion. Consequently, the proton concentration outside the membrane exceeds that inside, driving protons inward *via* diffusion through specialized channels. This proton influx facilitates ATP synthesis from ADP and Pi. Known as chemiosmosis, this process harnesses energy similar to osmosis to generate ATP.

• Utilizing Citric Acid Cycle Electrons for ATP Synthesis:

During glycolysis and pyruvate oxidation, NADH and $FADH_2$ molecules accumulate, each carrying a pair of electrons. These electrons are transferred to the cell membrane, where they engage with membrane-embedded proteins like NADH dehydrogenase. Afterward, electrons move through a sequence of respiratory proteins and carrier molecules referred to as the electron transport chain. In the final step of the chain, electrons react with oxygen molecules to produce water. This process pumps protons across the membrane, generating a proton gradient. Eukaryotes conduct oxidative metabolism within mitochondria, utilizing enzymes within the mitochondrial matrix for the citric acid cycle. Here, protons are pumped out of the matrix into outer compartments, creating a proton gradient. Protons re-enter the matrix through specialized channels, driving ATP synthesis by a large protein complex. Finally, ATP exits the mitochondrion and enters the cell's cytoplasm *via* facilitated diffusion, serving as a vital energy source.

Biology for Students, 2025, 31-51

CHAPTER 2

Cell Division (Cellular Reproduction): Exploring the Intricacies of Mitosis and Meiosis

Abstract: This chapter provides a detailed examination of the processes of mitosis and meiosis, crucial mechanisms in cellular reproduction. It begins with an overview of the nucleus and chromosomes, emphasizing their structure and function within the cell. The cell cycle is then dissected, highlighting its phasses and their significance in preparing a cell for division. Mitosis is explored through its phases—prophase, metaphase, anaphase, and telophase—focusing on the segregation of genetic material and the role of spindle microtubules. The chapter proceeds to discuss cytokinesis, comparing its occurrence in animal and plant cells. Meiosis is analyzed next, elucidating its role in sexual reproduction and the genetic diversity it fosters through crossing over and the production of haploid cells. The final section contrasts asexual and sexual reproduction, underscoring their respective advantages and disadvantages in terms of genetic diversity and adaptability.

Keywords: Cell cycle, Chromosomes, Cytokinesis, Meiosis, Mitosis.

UNDERSTANDING THE CHAPTER'S FOCUS: A ROADMAP

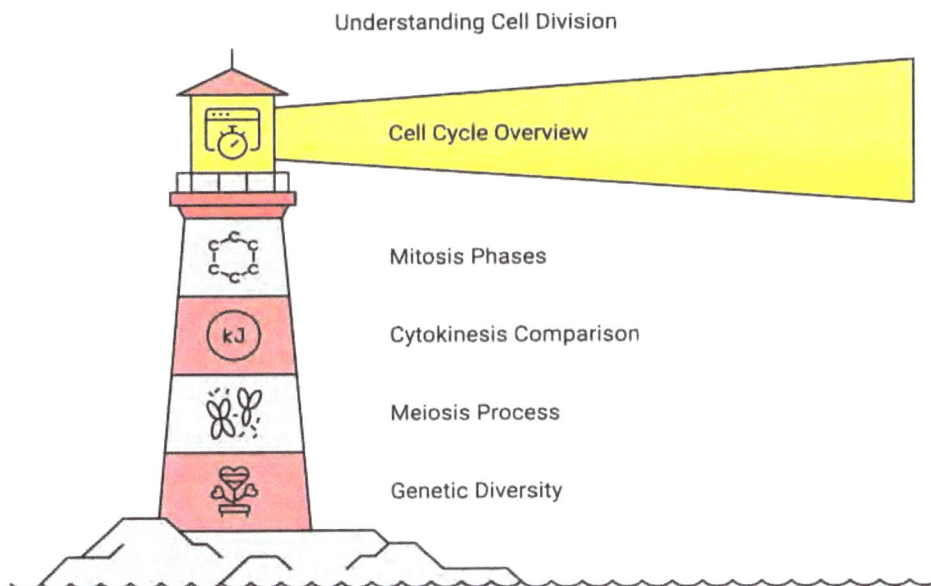

Understanding Cell Division

Cell Cycle Overview

Mitosis Phases

Cytokinesis Comparison

Meiosis Process

Genetic Diversity

Mohammad Mehdi Ommati

1. THE NUCLEUS AND CHROMOSOMES

The nucleus functions as the central repository of genetic information in a cell. Within the nucleus are chromosomes, composed of tightly coiled DNA strands interwoven with associated proteins. These proteins, known as **histones**, enable the DNA molecule to form bead-like complexes called **nucleosomes**. Further coiling and supercoiling result in densely packed structures known as chromosomes. **Chromatin**, the substance that constitutes chromosomes, is made up of long DNA strands bound with histones and nonhistone proteins.

A **karyotype** is a visual representation of an organism's chromosomes in their condensed and coiled state. It reveals that, apart from the sex chromosomes, chromosomes are organized in **homologous pairs**. Chromosomes other than sex chromosomes are referred to as autosomes. Cells containing two complete sets of parental chromosomes are called **diploid**, whereas those with only one set are **haploid**.

2. THE CELL CYCLE

The cell cycle describes the sequential events of cellular growth, preparation for division, and, eventually, division to produce two daughter cells, perpetuating the cycle. This cycle confers a form of immortality on single-celled organisms. In multicellular organisms, however, certain cell types, such as those in animal muscle and nerve tissues, slow down or exit the cycle entirely.

The conventional cell cycle comprises four distinct phases (Fig. **1**). The first three phases, collectively known as **interphase**, include:

- **G_1 phase**, characterized by normal metabolic activities;
- **S phase**, during which DNA replication and the synthesis of essential biological molecules occur; and
- **G_2 phase**, a short period of further growth and preparation for division.

The fourth phase, the M phase, encompasses mitosis, where replicated chromosomes condense and segregate, leading to cell division. The regulation of the cell cycle depends on the properties of the cytoplasm and external factors, such as stimulatory or inhibitory agents like **chalones**.

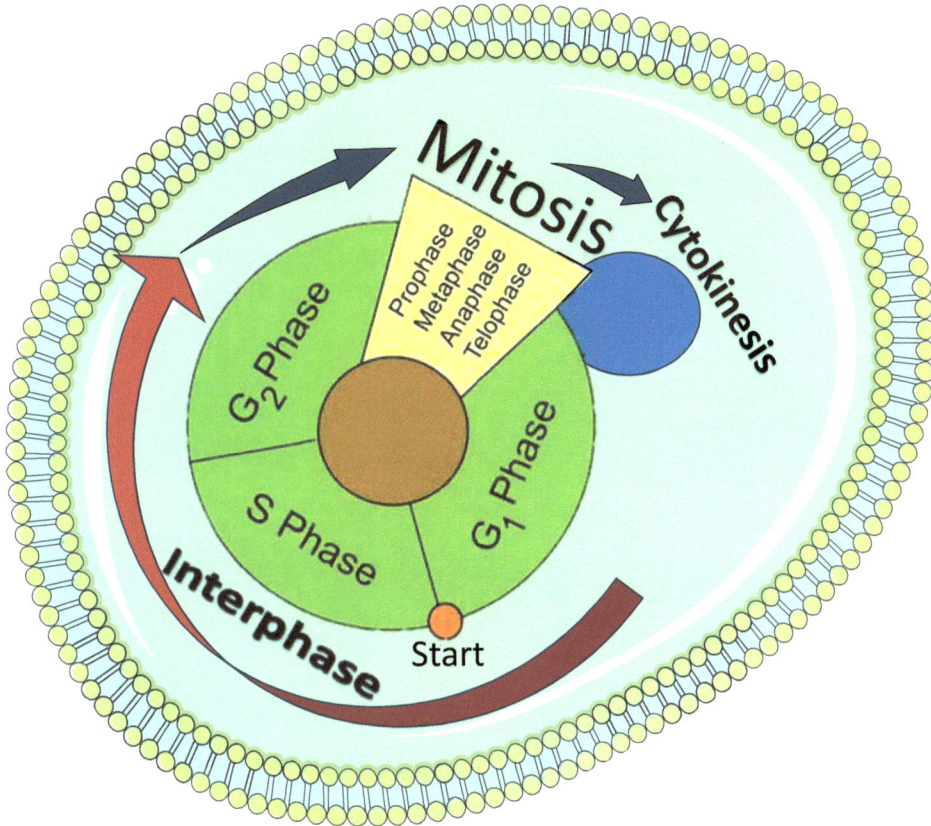

Fig. (1). Cell Cycle Stages: From Prophase to Cytokinesis.

3. MITOSIS: SEGREGATING THE GENETIC MATERIAL

Mitosis is divided into four distinct phases (Fig. **2**):

- **Prophase:** Chromosomes condense, each consisting of two highly compacted **chromatids** connected at a **centromere**.

- **Metaphase:** Chromosomes align along the spindle apparatus on the **metaphase plate**, a plane perpendicular to the **spindle** fibers.

- **Anaphase:** Chromatids of each chromosome separate and migrate toward opposite poles of the cell.

- **Telophase:** Nuclear envelopes form around the segregated chromosomes, and the cytoplasm divides.

Throughout mitosis, spindle microtubules play a pivotal role in ensuring the precise movement of chromatids. These microtubules extend from each pole of the dividing cell to the metaphase plate, forming the spindle apparatus. Additionally, during prophase, centromeric fibers extend from the spindle poles to specialized structures on chromosomes called kinetochores. Subsequently, during anaphase, these fibers contract, facilitating the separation of chromatids.

The formation of the spindle differs between animal and plant cells. In animal cells, it is associated with centrioles, while in plant and fungal cells, spindle formation occurs in regions known as microtubule organizing centers.

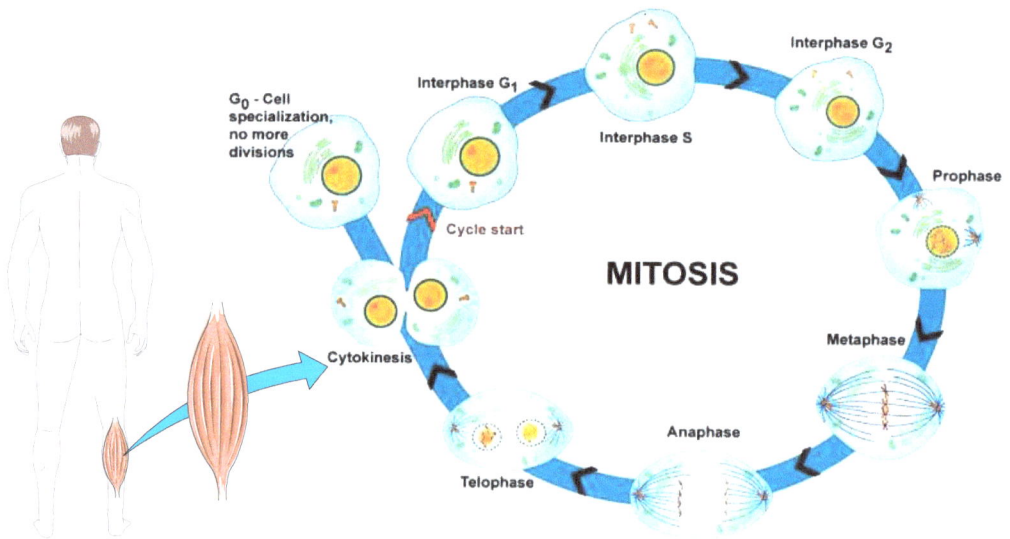

Fig. (2). Journey through mitosis into the muscular system: interphase to cytokinesis.

4. CYTOKINESIS: SEGREGATING THE CYTOPLASM

Cytokinesis, the final step of mitosis, involves the division of the cell cytoplasm. In animal cells, a ring of actin filaments contracts around the cell's equator, pinching the cell into two distinct daughter cells. In plant cells, however, the rigid cell wall necessitates the formation of a new **cell plate** at the cell's equatorial region. The cell wall material is subsequently deposited along the cell plate, completing the division.

5. MEIOSIS: FACILITATING SEXUAL REPRODUCTION

Meiosis is a specialized form of cell division that occurs in reproductive organs to produce sex cells (Fig. **3**). Like mitosis, meiosis begins with DNA replication but involves two successive nuclear divisions: meiosis I and meiosis II. These divisions result in four daughter cells, each with half the chromosome number of the parent cell. Notably, the process of crossing over during meiosis promotes genetic diversity by facilitating the exchange of genetic material between homologous chromosomes. This recombination ensures that the homologous chromosomes segregated into different progeny cells carry unique genetic combinations, enhancing genetic variability in sexually reproducing populations.

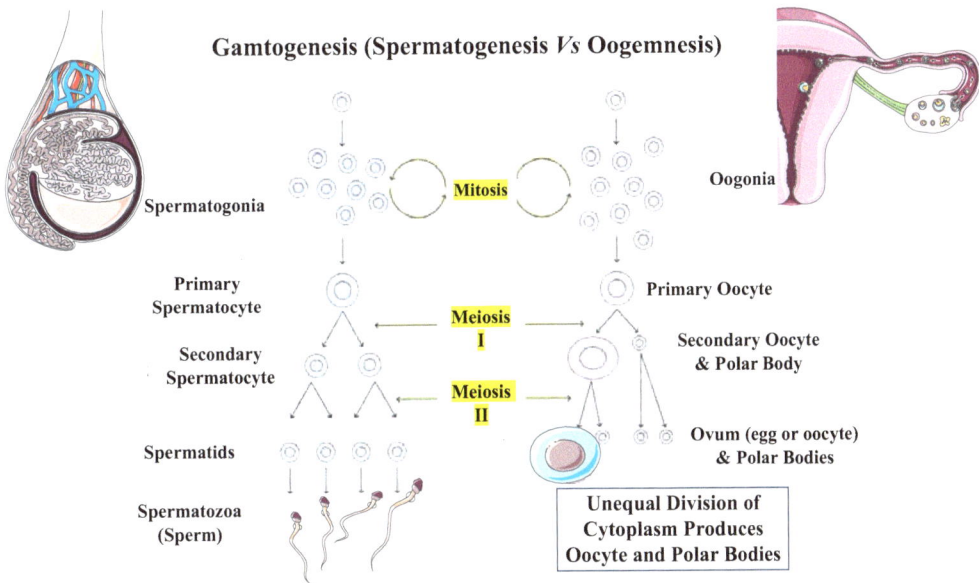

Fig. (3). Meiosis in progress, illustrating the formation of sperm cells.

In prophase I, homologous chromosomes pair through a process called synapsis, facilitated by the synaptonemal complex, a structure composed of protein and RNA. Unlike mitotic anaphase, during anaphase I, sister chromatids remain joined at the centromere and move together to one pole of the cell, halving the chromosome number in the daughter cells.

During telophase I, nuclear envelopes form around the chromosomes, and cytokinesis typically follows, completing the first division. Meiosis II begins with metaphase II, where chromosomes align on the metaphase plate, followed by centromere division and sister chromatids migrating to opposite poles. The process concludes with telophase II and a second round of cytokinesis, resulting in four haploid cells with randomly distributed parental chromosomes.

6. ASEXUAL *VERSUS* SEXUAL REPRODUCTION

Mitosis and meiosis underpin asexual and sexual reproduction, respectively, with each method offering distinct advantages.

- **Asexual reproduction** produces genetically identical offspring, preserving advantageous genetic traits, requiring minimal reproductive specialization, and proceeding more rapidly than sexual reproduction. However, this lack of genetic diversity renders populations more vulnerable to catastrophic events or diseases.
- **Sexual reproduction**, in contrast, enhances genetic diversity, mitigates deleterious mutations, and enables the emergence of novel gene combinations within populations.

Vocabulary (Index of Terms)

Anaphase [ˈænəˌfeɪz]

Anaphase is the third stage of mitosis, where sister chromatids separate as centromeres divide, and chromosomes migrate to opposite poles of the cell. This stage ensures equal distribution of genetic material to daughter cells, maintaining genomic integrity during cell division.

Cell plate [sɛl pleɪt]

The cell plate, unique to plant cells, forms during cytokinesis, starting from the center and extending to the cell membrane. It partitions cytoplasm and organelles between daughter cells, essential for their independent function and accurate cell division.

Chalone [ˈkæloʊn]

Chalones are substances discovered in mammalian tissue homogenates that, when administered to intact tissue cells, restrain mitosis, especially when adrenaline and corticosteroids are present. These molecules play a significant role in regulating cell division by exerting inhibitory effects, contributing to the maintenance of proper cell growth and proliferation. They act as a control mechanism, ensuring that cell division occurs only when necessary and in a regulated manner, thus preventing excessive or abnormal cell proliferation.

Chromatid [ˈkroʊməˌtɪd]	A chromatid is a replicated chromosome that remains physically connected to an identical copy at the centromere. This attachment ensures that each daughter cell receives a complete set of genetic information during cell division. Chromatids are essential for the accurate distribution of genetic material, ensuring that genetic information is faithfully transmitted to the next generation of cells.
Chromatin [ˈkroʊməˌtɪn]	Chromatin is the substance forming chromosomes, primarily composed of DNA and proteins, predominantly histones. It acts as a packaging material for genetic material in the cell nucleus, organizing DNA into compact structures. Chromatin's structure dynamically changes, affecting gene expression, DNA replication, and repair, crucial for regulating gene activity and maintaining genomic integrity.
Cytokinesis [ˌsaɪtoʊkɪˈniːsɪs]	Division of the cytoplasm of one cell into two new cells, marking the final stage of cell division, ensuring each daughter cell receives necessary cellular components for independent function.
Diploid [ˈdɪplɔɪd]	Diploid means having two sets of chromosomes, one from each parent.
Haploid [ˈheɪplɔɪd]	Haploid means having only one set of chromosomes instead of the usual two.
Histone [ˈhɪstoʊn]	Histones are simple proteins abundant in basic amino acids like arginine or lysine. They're closely associated with nucleic acids within the chromatin of eukaryotic cells. Histones play a crucial role in DNA packaging, organizing long strands of DNA into compact structures called nucleosomes, which are the basic units of chromatin. This packaging helps regulate gene expression and protects DNA from damage. Additionally, histones are involved in various cellular processes, including DNA replication, repair, and transcription.
Homologous pair [həˈmɑləgəs pɛr]	Homologous pairs are chromosomes that pair during meiosis, the process of cell division that produces reproductive cells. Each member of the pair, known as a homologue, is a duplicate of one of the chromosomes contributed during fertilization by either the mother or the father. Homologous chromosomes contain the same linear sequence of genes, and as a result, each gene is present in

duplicate. This redundancy allows for genetic variation through processes such as crossing over during meiosis, which promotes genetic diversity among offspring.

Interphase (cycle) ['ɪntər͵feɪz]   Interphase is the stage in the cell cycle where cells grow, replicate their DNA, and prepare for division.

Karyotype ['kæriə͵taɪp]   Karyotype refers to the visual appearance of the complete set of chromosomes in an organism or cell. It provides a detailed picture of the number, size, and shape of chromosomes, often used in genetics and medical diagnosis to identify chromosomal abnormalities or genetic disorders.

Meiosis [maɪ'oʊsɪs]   Meiosis is the process by which a cell nucleus divides into four daughter nuclei, each containing half the number of chromosomes found in the parent nucleus. It occurs in sexually reproducing organisms during the formation of gametes (sperm and eggs). Meiosis involves two successive divisions, resulting in the production of genetically diverse haploid cells, essential for sexual reproduction and the creation of genetic variation among offspring.

Metaphase ['mɛtə͵feɪz]   Metaphase is the second stage of mitosis, where chromosomes align along the equatorial plane of the cell. This alignment ensures that each daughter cell receives an equal and complete set of chromosomes during cell division. Metaphase is a highly regulated process, essential for accurate chromosome segregation and the proper distribution of genetic material to daughter cells.

Metaphase plate ['mɛtə͵feɪz pleɪt]   The metaphase plate is the arrangement of chromosomes in a single plane at the equator of the spindle during the metaphase stage of mitosis. This alignment ensures that each chromosome is properly positioned for separation during anaphase, ultimately leading to the accurate distribution of genetic material to daughter cells. The metaphase plate plays a crucial role in ensuring the fidelity of chromosome segregation during cell division.

Mitosis [maɪ'toʊsɪs]   Mitosis is a process that results in the equal and identical distribution of replicated chromosomes into two newly formed nuclei. It is a fundamental process in cell division, occurring in somatic cells to produce two daughter cells that are genetically identical to the parent cell. Mitosis consists of several stages, including prophase, metaphase, anaphase, and telophase, each with distinct events leading to the accurate segregation of chromosomes and the formation of two genetically identical daughter cells. This process plays a critical role in growth, development, and tissue repair in multicellular organisms.

Nucleosome ['njuːklɪə‚soʊm]

The nucleosome is the fundamental unit of chromatin structure in eukaryotic cells. It comprises eight histone molecules of four different types, around which approximately 140 base pairs of DNA are coiled. Nucleosomes play a crucial role in compacting and organizing DNA within the nucleus, forming the beads-on-a-string structure observed in chromatin. Additionally, nucleosomes regulate gene expression by controlling access to DNA, as the DNA wrapped around nucleosomes is less accessible to transcription factors and other regulatory proteins.

Prophase ['prəʊ‚feɪz]

Prophase is the initial stage of mitosis, characterized by the condensation and visibility of individual chromosomes. During prophase, the nuclear envelope dissolves, and the spindle apparatus begins to form.

Spindle ['spɪnd(ə)l]

The spindle is a structure composed of microtubules extending from pole to pole within the cell during cell division. It plays a vital role in the movement and segregation of chromosomes during both mitosis and meiosis.

Telophase ['telə‚feɪz]

Telophase is the final stage of mitosis, where daughter nuclei start to form. During telophase, the chromosomes reach the poles of the cell, and the nuclear envelope re-forms around them. Additionally, cytokinesis typically occurs during or shortly after telophase, completing the process of cell division.

COMPREHENSION TASKS

I. Key Terms: Matching

Instructions: Match each term from the left column with the corresponding description from the right column. Write the letter of the correct description next to the number of the term. Each description will only match with one term. Good luck!

1.	Cytokinesis	A.DNA+ histones
2.	Synapsis	B.Two sets
3.	Histone	C.X and Y
4.	Mitosis	D.One set
5.	Cell cycle	E.Chromosome display
6.	Chalone	F.Set of microtubules
7.	Spindle	G.Plant division
8.	Chromatid	H.Gamete production
9.	Nucleosome	I.Sequence of cell growth and division
10.	Diploid	J.Cell division
11.	Meiosis	K.Crossing over
12.	Cell plate	L.Inhibit cell division
13.	Sex chromosome	M.Division of cytoplasm
14.	Karyotype	N.Single chromosome copy
15.	Haploid	O.Positively charged protein

II. Cell Biology True or False Quiz: Determine whether the following statements are true or false based on your understanding of cell biology concepts. Write "True" if the statement is correct and "False" if the statement is incorrect. Good luck!

1.   ----- Autosomes encompass both the X and Y chromosomes.

2.   ----- Cells transition into the G_2 phase following the completion of the S phase.

3.   ----- Chalones facilitate cell division.

4.   ----- During telophase, the nuclear envelope reassembles.

5.   ----- Centrioles are most pronounced in animal cells.

6.   ----- Nuclear division may occur independently of cytokinesis.

7. ----- Synapsis involves the pairing of homologous chromosomes.

8. ----- Chromosomes undergo breakage at chiasmata.

9. ----- Cloning is not feasible for sexual organisms.

10. ----- All resulting daughter cells possess a haploid chromosome set.

III. Cell Biology Completion Exercise: Fill in the blanks with the appropriate term or phrase to complete each sentence related to cell biology. Write your answers in the space provided. Good luck!

1. When human chromosomes are stained on a slide, the resulting presentation, known as a ----------, should exhibit twenty-two pairs of ---------- alongside one pair of -------.

2. The S phase represents ----------. Because during this time, DNA is ---------- and histones are ----------.

3. In lieu of centrioles, most plants and fungi possess regions termed -------------- ------ centers.

4. During meiosis, unlike mitosis, homologous chromosomes undergo pairing through
----------.

5. Mitosis yields ----- identical offspring cells, each harboring a complete --------- - chromosome complement.

6. Meiosis generates ------- genetically diverse daughter cells, each containing a ---------- chromosome set.

7. Microtubules known as centromeric fibers bind to chromosomes at the --------- -----.

8. Two chromatids are linked at the --------------.

IV.   Multiple Choice: Choose the best answer for each question by circling the corresponding letter.

1. When does DNA replication occur?

  a. S phase

  b. M phase

c. G_2 phase

d. G_1 phase

e. G_0 phase

2.   Together, what do members of a chromosome pair form?

a. A tetrad

b. Chromatin

c. A homologous pair

d. A chromatid

e. A nucleosome

3.   What do autosomes represent?

a. All chromosomes in a normal human cell

b. Chromosomes found in egg or sperm cells

c. All chromosomes except the sex chromosomes

d. Chromosome pairs with unlike members

e. All homologous chromosomes

4.   Which phase is the longest in a typical vertebrate?

a.   S

b.   M

c.   G2

d.   G1

e.   None of the above

5. How many chromosomes does a normal diploid human cell contain?

a. 46 chromosomes

b. 23 chromosomes

c. 46 homologous pairs of chromosomes

d. 20 chromosomes

e. 20 pairs of chromosomes and 2 sex chromosomes

6. What can male sex chromosomes never be?

a. Haploid

b. Homologous

c. Diploid

d. Analogous

e. Duplicated

7. What phase is a cell in during G1, S, and G2?

a. The process of mitosis

b. The process of meiosis

c. Metaphase

d. Cytokinesis

e. Interphase

8. How can 2 meters of DNA fit into a human cell 5 micrometers in diameter?

a. DNA is broken into small fragments

b. DNA is wound around histones

c. DNA is wound around nonhistone proteins

d. Chromosomes are composed of chromatids

e. Chromosomes are joined at the centromere

9.   What is the association of a DNA molecule, histones, and nonhistone proteins known as?

a. Nucleosome

b. Chromosome

c. Chromatin

d. Chromatid

e. Karyotype

10.   What are the two daughter strands of a duplicated chromosome each known as?

a. Synapse pair

b. Centromere

c. Chromatid

d. Homologous chromosome

e. Chiasmata

11.   When do chromosomes become visible during the mitotic phase?

a. Metaphase

b. Anaphase

c. Prophase

d. Synapsis

e. None of the above

12. When does anaphase begin during mitosis?

a. Chromosomes line up in the nuclear region of the cell

b. Centromeres split and chromatids start to move apart

c. Synapsis occurs

d. Crossing over occurs

e. Prophase has been completed

13. When does a new nuclear envelope begin to form around each chromosome set?

a. Anaphase I

b. Metaphase

c. Prophase II

d. Cytokinesis

e. Telophase

14. What is spindle formation associated with in animal cells?

a. The nuclear membrane

b. Nucleosomes

c. Histone proteins

d. The centriole

e. Chromosomes

15.   What do haploid cells contain?

a. The diploid chromosome number

b. One copy of each chromosome

c. Two copies of the sex chromosomes

d. Twice the diploid chromosome number

e. Pairs of homologous chromosomes

16.   What does meiosis result in?

a. No change in the chromosome number

b. A doubling of the chromosome number

c. A reduction in the chromosome number

d. Two interphase cells

e. Four diploid cells

17.   When does crossing over occur?

a. G_0 phase

b. G_1 phase

c. Synapsis

d. Cytokinesis

e. M phase

18.   How is cytokinesis accomplished in animal cells?

a. A ring of actin filaments pinching the cell in two

b. The formation of a cell plate

c. The formation of a new cell wall

d. The action of centrioles

e. The re-formation of the nuclear membrane

19. What happens during crossing over?

a. Chromosomes line up along the metaphase plate

b. Homologous chromosomes exchange corresponding pieces of genetic information

c. Chromosomes move to opposite ends of the cell

d. The cell is in S phase

e. The cell is undergoing mitosis

20. What are the advantages of asexual reproduction?

a. It is more rapid than sexual reproduction

b. It requires few specialized reproductive structures

c. It preserves the individual's winning genetic makeup

d. All the above

e. a and c

READING UNDERSTANDING

The evolution of the central nervous system, comprising millions of nerve cells, can be traced back to the dispersed nerve cells of primitive aquatic organisms dating back half a billion years. The emergence of nerve cells is observed in coelenterates, such as hydra and sea anemones, which lacked centralized control in their nerve networks. This decentralized organization likely originated with flatworms, the earliest organisms with a distinct head region. While flatworms exhibit enhanced responsiveness to external stimuli through specialized sense cells, their behaviors primarily rely on instinct and reflexes due to the absence of a backbone. Notably, the evolution of intelligent behavior became feasible with the advent of larger and more complex brains, primarily observed in vertebrates. Despite the relatively larger brain size in fish compared to insects, the functional specialization of the fish brain underscores non-intellectual priorities, with distinct brain regions dedicated to olfaction, vision, and balance. Across mammalian evolution, brain complexity increased, with a shift in sense coordination towards the forebrain, particularly the cerebrum, facilitating memory and learning. Notably, in advanced mammals like primates (*e.g.*, monkeys, apes, and humans), adaptations to arboreal life led to a reduction in the olfactory region of the forebrain and an expansion of the visual processing regions, reflecting the increasing importance of vision in their ecological niche.

1. What is the primary focus of the passage?

A. The sensory organs of invertebrates

B. The anatomy of minute organisms

C. The genesis of the brain and central nervous system

D. The significance of vision for fish and advanced mammals

2. The hydra belongs to the category of ___ .

A. Flatworms

B. Coelenterates

C. Sea Anemones

D. Nerve Cells

3. In the final paragraph, the term "took to" could most aptly be substituted by which of the following?

A. Initiated

B. Migrated to

C. Adapted to

D. Became accustomed to

4. It can be deduced from the passage that insects lack __ .

A. Brains

B. Backbones

C. Nerve cells

D. Reflexes

5. According to the passage, what aids in orchestrating the intricate physical movements of a mammal?

A. The cerebellum

B. The forebrain

C. The cerebrum

D. The midbrain

6. What is the probable subject of the subsequent paragraph following the passage?

A. An elucidation of why certain animals inhabit trees

B. The differentiation between the brains of fish and primates

C. A comparative analysis of olfactory senses among different monkey species

D. The ongoing expansion of the mammalian brain

Reading Material: What Is Biology?

Biology, as a scientific discipline, encompasses the study of living organisms and their interactions with the environment. It relies on principles from chemistry and physics to elucidate the fundamental laws governing living systems. Given the diverse array of life forms, biology encompasses various specialized fields, ranging from practical applications like medicine, agriculture, and wildlife management to more theoretical domains such as medical microbiological physiology, photosynthetic biochemistry, and ethology. Additionally, there exists a facet of biology purely for enjoyment, such as insect collecting and bird watching.

The concept of life itself poses intriguing questions. Despite biology being defined as the study of living entities, delineating the essence of life remains a complex endeavor. This inquiry extends beyond theoretical realms, as legal frameworks necessitate precise definitions of life's onset and cessation, influencing matters like insurance claims and organ transplantation protocols. The definition of death, for instance, underscores the distinction between the cessation of the entire organism's functions and the demise of individual cells. Even in scenarios like heart transplants, where the donor may be legally declared deceased, the extracted organ retains vitality. Consequently, various forms of death exist, complicating the delineation of life.

Although a definitive definition of life eludes us, delineating its basic characteristics proves more feasible. Thus, rather than attempting a definitive definition, it is pertinent to outline the fundamental attributes of living organisms.

The Importance of Biology

Biological advancements have significantly contributed to our contemporary quality of life, particularly in two key areas: food production and disease management. Through selective breeding, plant and animal breeders have enhanced the productivity and quality of agricultural staples like corn, wheat, rice, and oats, resulting in increased food yields. Notably, modifications in farming practices, informed by biological experimentation, have further bolstered food production.

Similarly, animal breeders have achieved notable successes, with modern-day livestock exhibiting substantial improvements in productivity compared to their predecessors. For instance, poultry now lays more eggs, dairy cows yield greater milk quantities, and beef cattle grow at accelerated rates. Noteworthy transformations include the evolution of leaner pig breeds, tailored to meet shifting consumer demands away from fatty products.

Moreover, the effective control of pests and pathogens that threaten agricultural yields has been paramount in enhancing food security. Biologists' involvement in studying and managing these living threats has been instrumental in safeguarding agricultural productivity.

Beyond agriculture, remarkable strides have been made in healthcare and disease prevention. Vaccination programs have effectively curbed diseases like polio, measles, and mumps, significantly reducing their incidence. However, complacency regarding vaccination has led to a resurgence in some regions, underscoring the importance of sustained efforts in disease control.

Furthermore, advancements in understanding human physiology have facilitated the development of treatments for ailments such as diabetes, hypertension, and certain cancers, improving overall health outcomes. Paradoxically, these medical achievements contribute to a pressing biological challenge: the burgeoning global population.

Biological Challenges

Despite the numerous benefits conferred by biological advancements, certain challenges have arisen due to their misapplication or unintended consequences. For instance, conservation efforts aimed at preserving natural ecosystems have occasionally led to unforeseen ecological imbalances. The suppression of natural forest fires, intended to preserve trees, resulted in hazardous accumulations of combustible debris, ultimately exacerbating the severity of wildfires. Subsequent revisions in forest management policies, permitting controlled burns, underscore the complexities of balancing conservation goals with ecosystem health.

Additionally, the introduction of non-native species into new habitats has triggered ecological disruptions. Instances of foreign plant and animal species introduction, whether accidental or intentional, have often led to ecological havoc, with native species being outcompeted or driven to extinction.

Moreover, the ethical implications of biological advancements, particularly in healthcare, pose unresolved dilemmas. Technological breakthroughs enabling prolonged life have raised ethical questions concerning resource allocation and equitable access to healthcare. Disparities in healthcare provision, where affluent nations invest in cosmetic procedures while basic healthcare remains inaccessible to many, highlight ethical complexities in healthcare delivery.

In summary, while biology has yielded remarkable benefits in enhancing food security, disease control, and medical interventions, confronting the challenges arising from its misapplication and ethical implications remains imperative. A nuanced understanding of biological principles and their ethical ramifications is crucial for navigating these complexities and fostering sustainable advancements in biological sciences.

Section 2

Foundations of Genetics - History and Molecular Basis

CHAPTER 3

From Ancient Notions to Modern Insights: Unraveling the Foundations of Genetics

Abstract: This chapter traces the historical evolution of genetic theories from ancient concepts to contemporary understanding. It begins with early theories of inheritance, including Hippocrates' pangenesis and Weismann's germ plasm theory. The focus then shifts to Gregor Mendel, whose experiments with pea plants laid the groundwork for modern genetics. Mendel's laws of segregation and independent assortment are detailed, along with his methods and findings. The chapter concludes by discussing the rediscovery of Mendel's work and the chromosomal basis of heredity, highlighting contributions from scientists such as Sutton, Boveri, and Morgan. This comprehensive overview elucidates the foundational principles of genetics and their historical development.

Keywords: Alleles, Chromosomes, Inheritance, Mendel, Segregation.

UNDERSTANDING THE CHAPTER'S FOCUS: A ROADMAP

Evolution of Genetic Theories

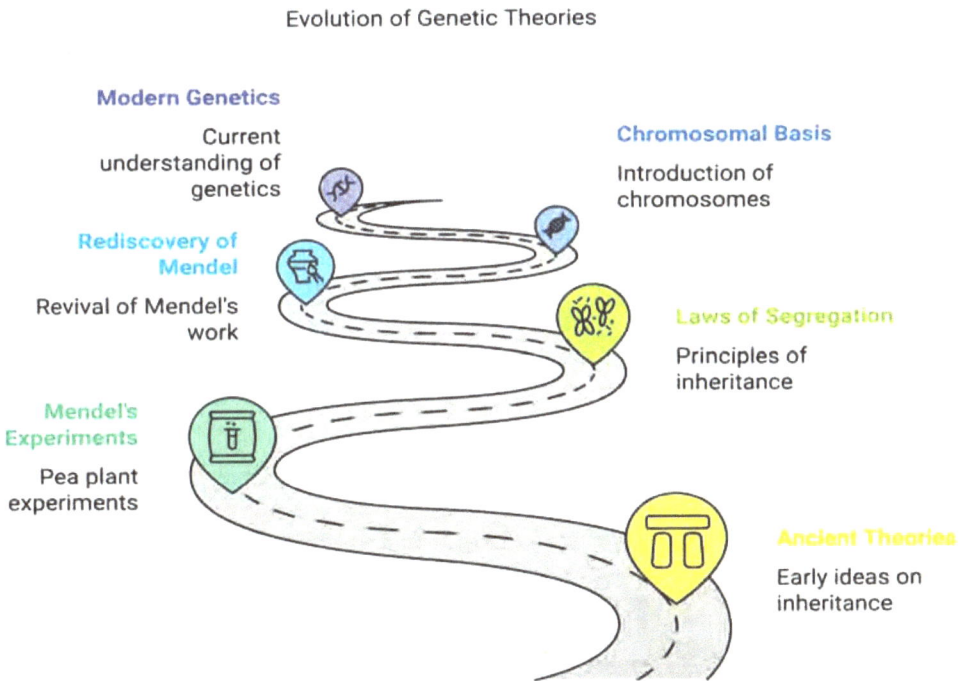

Mohammad Mehdi Ommati

1. EARLY THEORIES OF INHERITANCE

Early theories of inheritance, such as Hippocrates' **Pangenesis** and August Weismann's **Germ Plasm Theory**, laid crucial groundwork for understanding heredity. August Weismann (1834 – 1914), a German biologist, conducted experiments with mice and proposed that hereditary information is transmitted exclusively through gametes. His Germ Plasm Theory suggested that "germ cells" (sperm and egg cells) carry hereditary material, while somatic cells do not. This was a significant departure from earlier notions, like the blending hypothesis, which posited that traits from both parents combine in offspring, resulting in a loss of distinct characteristics. Weismann demonstrated that acquired traits, such as the removal of a mouse's tail, do not influence hereditary traits passed to offspring. His findings refuted the concept of inheritance of acquired traits, strengthening the idea that germ cells are the sole carriers of genetic information.

Pangenesis, attributed to ancient Greek philosopher Democritus (c. 460[1] – c. 370[2] BCE), proposed that particles from different parts of the body (pangenes) contribute to the formation of offspring. While Hippocrates (c. 460 – c. 375 BCE), often referred to as the "Father of Medicine," made notable contributions to medical science, there is no direct evidence to suggest he explicitly rejected pangenesis. It is important to note that pangenesis was later disproven by scientific discoveries, such as Gregor Mendel's (1822 – 1884) groundbreaking work in genetics, which provided a more accurate understanding of heredity.

On the other hand, Hippocrates, often referred to as the "Father of Medicine," lived in ancient Greece and made significant contributions to medical science. While Hippocrates made many observations about health and disease, there is no direct evidence to suggest that he specifically rejected the idea of pangenesis. Pangenesis was a theory proposed by the ancient Greek philosopher Democritus, who lived before Hippocrates. According to pangenesis, particles called pangenes from different parts of the body contribute to the formation of offspring. However, this theory was later refuted by later scientific discoveries, such as Mendel's work on genetics in the 19th century, which provided a more accurate understanding of heredity. Therefore, it is unlikely that Hippocrates explicitly rejected pangenesis since it was not a widely held or established concept during his time.

---

[1] c. 460 BCE means approximately 460 years before the Common Era (or before the traditional birth of Christ).

[2] c. 370 BCE means approximately 370 years before the Common Era.

2. GREGOR MENDEL: PIONEER OF GENETIC INQUIRY

Gregor Mendel, often regarded as the "Father of Genetics," revolutionized our understanding of inheritance. As an Augustinian monk in Brünn, Austria, Mendel combined his mathematical training with a keen interest in natural science. His experiments, designed to investigate the particulate nature of heredity, challenged the prevailing blending hypothesis. Despite the groundbreaking nature of his findings, Mendel's work was not fully appreciated until decades after his death.

3. MENDEL'S SEMINAL INVESTIGATIONS

Mendel's pioneering investigations focused on plant breeding experiments with the garden pea, a species characterized by self-fertilization and true breeding. Selecting seven distinct traits, such as seed color and plant height, each with clear binary outcomes, Mendel meticulously recorded progeny characteristics across two generations to challenge the prevailing blending theory (Fig. **1**).

In his experiments, Mendel consistently observed **dominant** and recessive traits, with a 3:1 ratio of dominant to **recessive** traits in the second filial (F2) generation. Mendel proposed that organisms inherit two hereditary units for each trait, one from each parent. These units, now known as **alleles**, are alternative forms of genes—the basic units of heredity. He clarified that organisms with identical alleles (*e.g.*, TT or tt) are **homozygous**, while those with differing alleles (*e.g.*, Tt) are **heterozygous**. Importantly, dominant alleles express their traits in the phenotype, while recessive alleles do not manifest unless paired together.

In heterozygous organisms, the dominant allele typically manifests in the **phenotype**, representing its physical appearance, while the **genotype** encompasses both the dominant and recessive alleles, reflecting the genetic composition. To visualize the possible allele pairings from a genetic cross, Mendel employed the **Punnett square**, which illustrates the potential genotypes and phenotypes of the offspring. Mendel's experiments on dominant and recessive inheritance led to his first law, the **law of segregation**. This principle states that each organism inherits one allele from each parent, forming an allele pair, and during meiosis, these alleles separate into different gametes. Mendel performed **test crosses** to confirm his theory, mating organisms of unknown genotype with those homozygous recessives for the trait in question, observing the ratio of dominant phenotypes in the offspring to determine the genotype of the unknown parent.

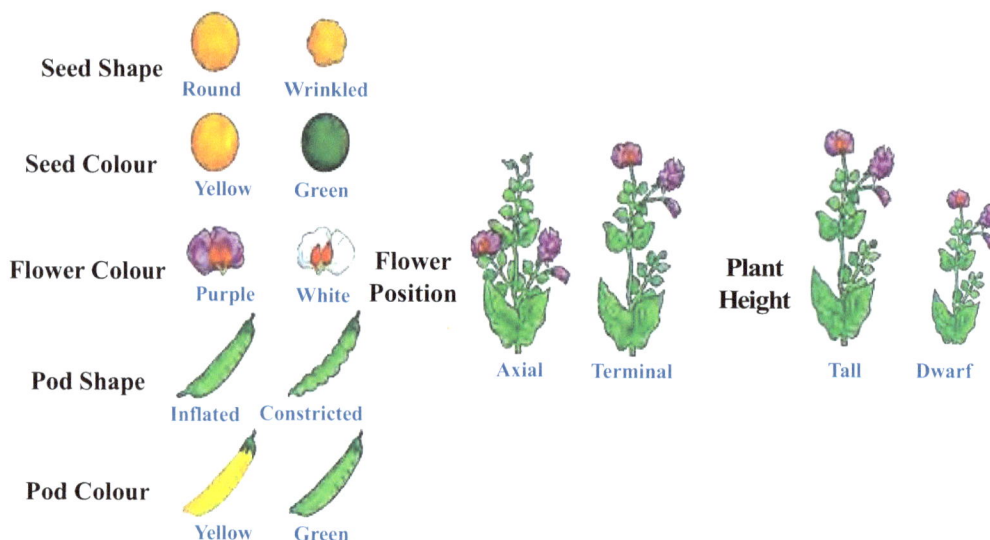

Fig. (1). Mendel's seven pairs of contrasting traits in pea plants.

4. MENDEL'S INSIGHTS AND THE PRINCIPLE OF INDEPENDENT ASSORTMENT

By performing **dihybrid crosses**, Mendel extended his research to examine how two traits are inherited simultaneously. He observed that the inheritance of one trait does not influence the inheritance of another, leading to **the principle (law) of independent assortment**. This principle asserts that alleles of different genes assort independently during gamete formation, though exceptions like **incomplete dominance** exist. In such cases, the phenotype reflects a blend of both alleles' effects, but the alleles themselves remain discrete.

5. MENDEL'S CONTRIBUTIONS AND REDISCOVERY

Mendel published his revolutionary findings in 1866 through the Brünn Society for Natural History. However, his work was largely ignored until 1900, when Carl Correns, Hugo de Vries, and Erich von Tschermak independently rediscovered his experiments. This rediscovery ushered in a new era of genetic research.

6. CHROMOSOMES AND THE FOUNDATIONS OF MENDELIAN GENETICS

The chromosomal basis of Mendel's theories was proposed by Walter Sutton and Theodor Boveri, who independently hypothesized that hereditary units reside on chromosomes. Morgan's research on fruit flies validated Mendelian principles while uncovering phenomena like sex-linked inheritance. His collaborator, Calvin Bridges, later discovered **nondisjunction**, a failure of chromosomes to segregate properly during meiosis, further advancing the understanding of genetic inheritance.

Vocabulary (Index of Terms)

Allele [əˈliːl]	An allele refers to the alternative form of a gene that occupies the same locus or position on a chromosome. For instance, in the case of ear shape, the attached earlobe allele and the free earlobe allele represent different variations in the gene responsible for determining ear shape. These alleles can produce different phenotypic expressions of the same trait.
Dihybrid cross [daɪˈhaɪbrɪd krɒs]	A dihybrid cross involves the mating of organisms that exhibit contrasting traits for two specific gene pairs. This breeding experiment allows for the examination of the inheritance patterns of two different traits simultaneously.
Dominant [ˈdɒmɪnənt]	The dominant allele in a gene pair manifests its effect on the phenotype regardless of the presence of another allele. This allele typically produces a visible trait in the organism's phenotype when present in either homozygous dominant or heterozygous genotype.
Gene [dʒiːn]	A gene is a hereditary unit located on a chromosome, composed of a sequence of DNA nucleotides.
Genotype [ˈdʒiːnəˌtaɪp]	The genotype refers to the complete set of genes present in an organism, encompassing both expressed and non-expressed genes.
Germ plasm theory [dʒɜːrm plæzəm ˈθɪəri]	The germ plasm theory posits that a substance, presumed to be transmitted through the gametes (germ cells), remains unchanged from one generation to the next. It was believed that this germplasm was unaffected by

environmental factors and was responsible for generating the body cells.

Heterozygous [ˌhɛtəroʊˈzaɪgəs]

A heterozygous organism refers to a diploid individual possessing two different allelic forms of a specific gene, one inherited from each parent.

Homozygous [ˌhoʊmoʊˈzaɪgəs]

Homozygous describes a diploid organism having two identical alleles for a particular trait. These alleles can be either dominant or recessive, resulting in uniform expression of the associated phenotype.

Incomplete dominance [ˌɪnkəmˈpliːt ˈdɒmɪnəns]

Incomplete dominance occurs when two allelic genes exhibit a phenotype in the heterozygous state that is intermediate between the phenotypes of the homozygous states for each allele. This phenomenon contrasts with complete dominance, where one allele completely masks the expression of the other.

Law of independent assortment [ˈlɔː əv ˌɪndɪˈpɛndənt əˈsɔːrtmənt]

The law of independent assortment states that during gamete formation, the alleles of different gene pairs segregate independently of each other. This means that the inheritance of one gene pair is not influenced by the inheritance of another gene pair. This principle is a fundamental aspect of Mendelian genetics and contributes to genetic variation in offspring.

Law of segregation /ˌsɛgrɪˈgeɪʃən/

When gametes are formed by a diploid organism, the alleles that control a trait sepa-rate from one another into different gametes, retaining their individuality.

Nondisjunction [nɒndɪsˈdʒʌŋkʃən]

Nondisjunction refers to the failure of paired chromosomes to separate during metaphase, leading to one daughter cell receiving both chromosomes and the other receiving none of the chromosomes involved. This can happen during either meiotic or mitotic cell divisions.

Pangenesis [pænˈdʒenəsɪs]

Pangenesis is a theory of heredity proposing that tiny particles called "gemmules," "humours," or "essences" migrate from various body cells to the reproductive organs, where they contribute to the formation of gametes. These gametes, thus, carry the hereditary information from all parts of the body to the next generation.

Phenotype [ˈfiːnəˌtaɪp]

Phenotype refers to the observable characteristics of an organism, encompassing its physical, chemical, and sometimes psychological traits, resulting from the expression of its genes and the interaction with the environment.

Punnett square ['pʌnɪt skwɛə]

A Punnett square is a graphical method used to predict the probabilities of different genotypic combinations in offspring based on the alleles contributed by the parents.

Recessive [rɪ'sɛsɪv]

A recessive allele is one of a pair of alleles that only manifests its effect in the phenotype when the individual is homozygous for that allele, as it is masked by the presence of a dominant allele in a heterozygous genotype.

Test cross [test krɒs]

A test cross is a breeding experiment performed to determine the genotype of an organism with an unknown genotype for a particular trait. It involves crossing the organism with a homozygous recessive individual for the trait in question. This allows for the determination of whether the unknown organism is heterozygous or homozygous dominant for the trait based on the phenotypic ratios observed in the offspring.

COMPREHENSION TASKS

I. Key Terms: Matching

Instructions: Match each term from the left column with the corresponding description from the right column. Write the letter of the correct description next to the number of the term. Each description will only match with one term. Good luck!

1. Dominant		**A.** Hippocrates
2. Phenotype		**B.** Parental
3. Allele		**C.** Nondominant
4. Homozygous		**D.** Always expressed
5. Recessive		**E.** Alternative forms
6. P1		**F.** Grandchildren
7. Dihybrid cross		**G.** Basic units of heredity
8. Law of segregation		**H.** Different
9. Gene		**I.** Similar
10. Pangenesis		**J.** Weismann
11. F2		**K.** Total alleles
12. Sex-linked		**L.** Two characters
13. Heterozygote		**M.** Separate
14. Genotype		**N.** X or Y chromosome
15. Germplasm theory		**O.** Appearance

II. Cell Biology True or False Quiz: Determine whether the following statements are true or false based on your understanding of cell biology concepts. Write "True" if the statement is correct and "False" if the statement is incorrect. Good luck!

1. ----- The Germ Plasm Theory suggests that each part of the body generates a "seed".

2. ----- When offspring "breed true", it means they exhibit identical traits to their parents.

3. ----- Alleles represent alternative versions of genes.

4. ----- The color inheritance in snapdragons exemplifies blending inheritance.

5. ----- Nondisjunction can lead to the emergence of an XO male.

6. ----- Hippocrates completely dismissed the concept of pangenesis.

7. ----- Organisms that are homozygous produce gametes of only one gene type.

8. ----- Test crosses were instrumental in establishing the Law of Segregation.

9. ----- Sutton and Boveri formulated the Chromosome Theory of Heredity.

10. ----- Mendel's second law asserts that the inheritance of traits is dependent on each other.

III. **Completion Exercise: Complete each sentence by filling in the blanks with the appropriate term or phrase**. Write your answers in the space provided. Good luck!

1. The genetic phenomenon of _____ produces intermediate phenotypes.

2. The discovery that the units of heredity reside on chromosomes is credited to _____.

3. Mendel elucidated his second law through a series of _____ experiments with his pea plants.

4. Although Mendel's contributions went unrecognized during his lifetime, they were later acknowledged by ___.

5. A _____ provides a simplified grid structure facilitating genetic cross-calculations involving multiple traits.

6. _____ represent alternative forms of the same gene.

7. Mendel found peas to be ideal for selecting true-breeding strains due to their _____ characteristics.

IV. **Multiple Choice:** Choose the best answer for each question by circling the corresponding letter.

1. The concept that each part of an organism's body produces a "seed" that then moves to the reproductive organs is known as, _____.

a. Germplasm theory

b. Pangenesis

c. Law of independent assortment

d. Law of segregation

e. Heterozygosity

2. The Germplasm theory of heredity, which emphasizes the significance of parental germ cells in determining traits in offspring, contradicted prevailing ideas of inheritance, particularly _____.

a. Blending inheritance

b. Vitalism

c. Pangenesis

d. Mendelism

e. Herodotus

3. When true-breeding plants undergo self-fertilization, they _____.

a. Produce heterozygous offspring

b. Produce offspring that exhibit a range of phenotypes

c. Produce offspring that are identical to the parent

d. Produce offspring with dominant genotypes

e. Produce offspring with recessive genotypes

4. Mendel's experiments with peas had distinctive and significant characteristics, including the fact that _____.

a. He studied traits that offered only two alternative outcomes

b. He meticulously documented the type and quantity of all offspring

c. He tracked the outcomes of each cross for two generations

d. All of the above

e. He utilized the test cross procedure for two generations

5. A plant with differing alleles for a particular trait _____ .

a. Possesses identical alleles for that trait

b. Bears dissimilar alleles for that trait

c. Can be distinguished from a plant with homozygous alleles by its phenotype

d. Exhibits incomplete dominance

e. Does not fit any of the above descriptions

6. In contemporary terms, Mendel's primary principle indicates that _____ .

a. Allele segregation occurs randomly during meiosis

b. Homologous chromosomes pair up during meiosis

c. Chromosomes migrate in a predetermined manner

d. Recessive traits manifest exclusively in homozygous individuals

e. Dominant traits manifest exclusively in homozygous individuals

7. Traits governing green and yellow seed color exemplify _____ .

a. Incomplete dominance

b. Distinct alleles of the same gene

c. Linkage to sex chromosomes

d. The interaction of two distinct genes

e. Involvement of sex chromosomes

8. Assuming P represents a dominant allele and p a recessive allele for a given trait, a homozygous recessive individual will have the following genotype: _____ .

a. PP

b. Pp

C. pP

d. pp

e. none of the above

9. According to Mendel's principles, in a cross between a homozygous dominant individual and a homozygous recessive individual for the same trait, _____ .

a. The recessive trait will appear in a 1:3 ratio among the F1 phenotypes

b. The recessive trait will not be observed in the F1 phenotypes

c. The recessive trait will be more prevalent in the F1 phenotypes compared to the dominant trait

d. The recessive trait will appear three times as often as the dominant trait in the F1 phenotypes

10. Mendel's principle of independent assortment asserts that _____ .

a. Monohybrid crosses will exhibit segregation and independent assortment

b. The separation of alleles of one gene relies on the separation of all other alleles

c. Test crosses consistently yield heterozygous offspring

d. Alleles of different genes segregate randomly, and fertilization occurs at random

e. Alleles separate during meiosis

11. In a cross between a homozygous plant bearing round yellow seeds and a homozygous plant bearing wrinkled green seeds, all the offspring had round yellow seeds. If R represents the round allele and Y the yellow allele, the F 1 genotype was _____ .

a. RRYY

b. RRyy

c. RRyY

d. RrYy

e. Rryy

12. Mendel employed the Punnett square to_____ .

a. Forecast the likelihoods of various allele combinations

b. Document the outcomes of his test crosses

c. Develop the concept of the dihybrid cross

d. Refute the idea of pangenesis

13. A dihybrid cross involves mating between two organisms ____ .

a. When at least one carries two different traits

b. When both carry two different traits

c. When one is homozygous and the other is heterozygous for a trait

d. When both are homozygous for a single trait

14. In certain cases, offspring phenotypes are intermediate between those of their parents. This is an instance of ___ .

a. Blending

b. Linkage

c. Incomplete dominance

d. Intermediate inheritance

15. Sutton and Boveri postulated that Mendel's hereditary factors were situated on chromosomes, in a theory known as ___ .

a. Pangenesis

b. Germplasm theory

c. Chromosomal theory of inheritance

d. Independent assortment

e. Blending theory of inheritance

16. The first discovery of sex chromosomes occurred in___ .

a. Peas

b. Mame cats

c. Frizzle chickens

d. Fruit flies

e. Humans

17. Sex-linked traits are ____ .

a. Traits located on sex chromosomes

b. Traits exclusive to one sex

c. Traits observed in female fruit flies

d. Traits carried only on the Y chromosome

e. None of the above

18. Nondisjunction describes_____ .

a. The movement of chromosomes to opposite poles of the cell during anaphase

b. The failure of homologous chromosomes to segregate during mitosis or meiosis

c. The failure of two new daughter cells to separate after mitosis
d. The process of crossing over during meiosis

e. None of the above

19. In fruit flies, the following genotypes will yield a female ___.

a. YY

b. YO

c. XY

d. XO

e. XX

20. A gamete with a novel combination of alleles, differing from the parental associations, is known as ___.

a. A recombinant type

b. A crossing-over type

c. An incomplete dominant

d. A heterozygote

e. A homozygote

READING UNDERSTANDING

Insects face numerous threats and have short lifespans, yet they must endure long enough to reproduce and ensure the survival of their species. Resembling or imitating plants to appear "inedible" is a widespread strategy among insects, enhancing their chances of survival. While mammals rarely employ such camouflage, many fishes and invertebrates do.

The stick caterpillar aptly earns its name by closely resembling a brown or green twig, making it difficult to discern. This common caterpillar can be found across North America and is also known as the "measuring worm" or "inchworm." Its unique mode of locomotion involves arching its body, grasping branches with its front feet, and looping its body forward. When threatened, the stick caterpillar extends its body away from the branch at an angle, mimicking a twig, until the danger passes.

Walkingsticks, or stick insects, achieve protection without assuming a rigid, twig-like posture; they naturally resemble inedible twigs in any position. With various species ranging from a few inches in North America to tropical varieties exceeding a foot in length, these insects have front legs extended when at rest. Some tropical species mimic thorny bushes or trees with spines or ridges, further enhancing their camouflage.

Leaves serve as another favored object for insect imitation. Many butterflies, for instance, can vanish by folding their wings and quietly blending into the foliage they resemble, providing them with effective concealment.

1. What is the primary focus of the passage?

A. Caterpillars inhabiting trees

B. Insect feeding behaviors

C. Mechanisms of insect camouflage

D. Endangered insect species

2. In line 1, the term "enemies" refers to ___.

A. Competing creatures for space

B. Severe weather conditions

C. Predatory creatures consuming insects

D. Inedible insects

3. How does the stick caterpillar achieve a twig-like appearance, as per the passage?

A. By maintaining a rigid and motionless posture

B. By coiling around a twig

C. By altering the color of its exoskeleton

D. By flattening its body against a branch

4.   Which statement accurately characterizes stick insects?

A. They continually blend with their environment.

B. They mimic other insects.

C. They exhibit camouflage solely while moving.

D. They alter their color to become invisible.

5.   Which objects are NOT cited in the passage as imitated for protective purposes?

A. Thorns

B. Flowers

C. Leaves

D. Sticks

6.   In which paragraph does the author delineate the locomotion pattern of stick caterpillars?

A. Paragraph one

B. Paragraph two

C. Paragraph three

D. Paragraph four

Reading Material: Genes and Traits

In every diploid organism, there exists a pair of genes for each characteristic, referred to as alleles. Within the population, multiple alternative forms of these genes may exist. For instance, humans possess two alleles dictating earlobe shape: one yields a free-hanging fleshy lobe, while the other results in an attached lobe. The expression of earlobe type is determined by the alleles inherited from each parent and their interaction. These alleles consistently reside on homologous chromosome pairs, one per chromosome, in a species. This positioning remains consistent across individuals within a species.

The genome encompasses all genes necessary to define an organism's complete array of traits. A diploid (2n) cell carries two genomes, while a haploid cell (n) holds one. An organism's genotype represents a compilation of its genes, housed within the cell's DNA. Although an organism's complete genotype is often inaccessible, determining the genes responsible for particular traits is frequently feasible. For example, humans harbor two earlobe-shaped alleles, allowing for three potential genotypic combinations. These combinations correspond to distinct phenotypes of earlobe appearance.

The manner in which gene combinations manifest is referred to as an organism's phenotype. Individuals with varying genotypes may exhibit identical phenotypes. External factors, such as the environment, can influence gene expression. Certain genes may remain dormant until specific environmental conditions are met. For instance, cats may display coat-color genes exclusively when skin temperature drops, resulting in distinct coloration on their ears and paws. Similarly, human genes for freckles may require sunlight exposure to fully express.

The interaction between alleles also shapes characteristics. Homozygous individuals possess identical alleles for a trait, leading to consistent expression of that trait. Conversely, heterozygous individuals harbor different allelic forms of a gene, resulting in a blended expression. Dominant alleles overpower recessive ones in expression, but exceptions exist, with variations in dominant gene manifestation. Environmental factors, both internal and external, further influence gene expression. For instance, hormonal changes during puberty can alter voice pitch, while dietary adjustments can delay the onset of genetic disorders like diabetes mellitus.

CHAPTER 4

The Chemical Foundations and Structural Unveiling of the Gene

Abstract: This chapter delves into the chemical and structural basis of genes, starting with early investigations that linked genes to enzyme activity. It highlights significant milestones, such as Beadle and Tatum's one-gene-one-enzyme hypothesis and Linus Pauling's refinement to one-gene-one-polypeptide. The chapter also explores the discovery of DNA's chemical composition, emphasizing contributions from scientists like Miescher, Avery, and Levene. Chargaff's rules and the race to unveil DNA's double-helix structure by Watson and Crick are discussed. Finally, it explains the mechanisms of DNA replication, emphasizing the semiconservative model and the roles of leading and lagging strands.

Keywords: Base pairing, DNA replication, DNA structure, Gene expression, Nucleic acids.

UNDERSTANDING THE CHAPTER'S FOCUS: A ROADMAP

The Unraveling of DNA and Genetic Understanding

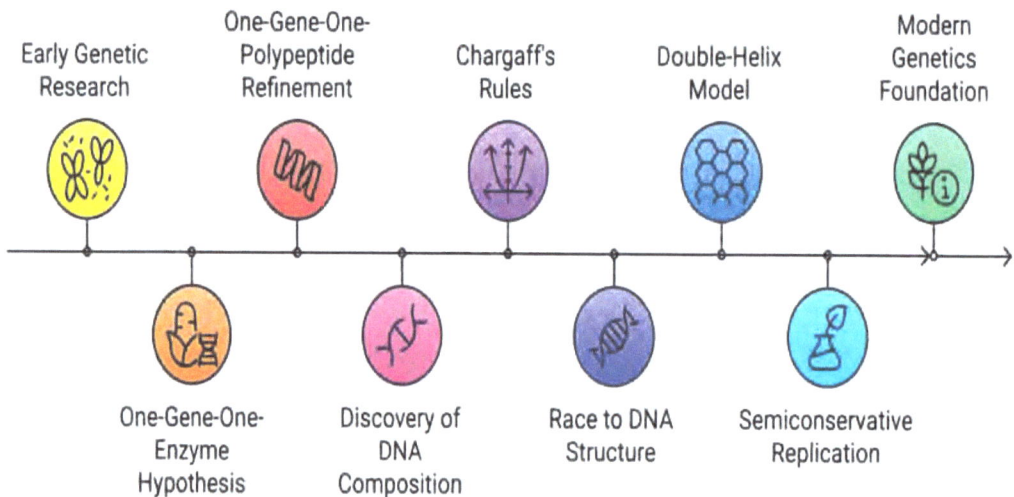

Mohammad Mehdi Ommati

1. EXPLORING THE CHEMICAL BASIS OF GENETIC EXPRESSION

Sir Archibald Garrod emerged as a pioneering figure in linking genes to phenotypic traits, particularly through his early work on alkaptonuria, which suggested a correlation between genes and enzyme activity. Building on Garrod's insights, Beadle and Ephrussi conducted seminal studies demonstrating the association between specific genes and biosynthetic processes governing eye pigmentation in *Drosophila melanogaster*. Subsequently, Beadle and Tatum's landmark investigations into mutation effects in *Neurospora crassa* led to the formulation and exploration of the **one-gene-one-enzyme hypothesis**, positing a direct correspondence between individual genes and specific enzymatic functions. Their groundbreaking research laid the groundwork for subsequent inquiries into the intricate mechanisms through which enzymes regulate complex metabolic pathways. Furthermore, Linus Pauling's investigations into the role of hemoglobin in sickle cell anemia contributed to refining the initial hypothesis, culminating in the proposition of the **one-gene-one-polypeptide paradigm (hypothesis)**.

Neurospora crassa, a model organism widely employed in genetic research, particularly in the study of mutants exhibiting auxotrophic traits, has played a pivotal role in elucidating genetic pathways and metabolic processes.

2. PURSUING THE CHEMICAL COMPOSITION AND STRUCTURAL CHARACTERISTICS OF NUCLEIC ACIDS

Johann Miescher's isolation of nucleic acids in 1871 was a pivotal step in unraveling the molecular constituents of chromosomes, further advanced by Feulgen's red-staining technique for chromosomal visualization. Frederick Griffith's seminal experiments with pneumococcal strains underscored the transformative potential of unidentified bacterial material in phenotypic alteration, laying the groundwork for subsequent revelations regarding the nature of this transformative substance. Building upon Griffith's findings, Avery, MacLeod, and McCarty definitively identified this elusive material as deoxyribonucleic acid (DNA) in the 1940s. Concurrently, P.A. Levene's elucidation of DNA's structural components, comprising four nitrogenous bases—adenine, guanine, cytosine, and thymine—each bound to a sugar-phosphate backbone termed nucleotides, provided critical insights into DNA's molecular architecture. Martha Chase and Alfred Hershey's pivotal experiments (in the early 1950s) with Escherichia coli decisively resolved the debate regarding the carrier of genetic information, affirming DNA's central role in heredity. The structural elucidation of DNA revealed a double-helix configuration, wherein nucleotide base pairs—adenine-thymine and cytosine-guanine—are interconnected by hydrogen bonds, thereby establishing the foundation for genetic encoding within the DNA molecule. Within the structure of each DNA nucleotide lies a fundamental composition consisting of a pentose sugar (a five-carbon sugar), specifically deoxyribose, which is covalently linked to one of four nitrogenous bases: **adenine**, **guanine**, **cytosine**, or **thymine**. Notably, adenine and guanine have a distinctive molecular structure with double-ring structures, classifying them as **purines**, while cytosine and thymine have single-ring structures, categorizing them as **pyrimidines**. The amalgamation of a nitrogenous base with the deoxyribose sugar forms a foundational unit termed a **nucleoside**, representing the fundamental building block of DNA's intricate molecular framework. Chargaff's rules delineate two fundamental observations regarding the composition of DNA. Firstly, they stipulate a consistent parity between the quantities of adenine and thymine bases, as well as between cytosine and guanine bases, within the DNA molecule. Specifically, the principle asserts that the abundance of adenine is equivalent to that of thymine, while the quantity of cytosine matches that of guanine, thereby maintaining a balanced nucleotide composition critical for DNA stability and function.

Secondly, Chargaff's rules highlight interspecies variations in the ratios of adenine-thymine (A-T) and cytosine-guanine (C-G) base pairs. These relative proportions can differ across diverse organisms, reflecting the species-specific genomic signatures encoded within the DNA sequences. Thus, while the underlying principles of base pairing remain consistent, the precise ratios of complementary base pairs may exhibit variability, underscoring the adaptability and evolutionary divergence inherent in DNA structures among different biological taxa.

3. THE RACE FOR UNRAVELING THE MOLECULAR STRUCTURE OF DNA

In the quest to elucidate DNA's structure, researchers in the late 1940s and early 1950s drew on a confluence of scientific insights, including Chargaff's observations, Levene's biochemical analyses, and the pivotal **X-ray diffraction** images captured by Rosalind Franklin and Maurice Wilkins. These images provided crucial evidence for DNA's helical structure, informing Watson and Crick's double-helix model. Drawing on these multidisciplinary inputs, Watson and Crick proposed the iconic **double-helix model** of DNA, which described a helical framework consisting of intertwined sugar-phosphate backbones and nucleotide base pairs. Central to their model was the concept of complementary base pairing—adenine with thymine (A-T), and cytosine with guanine (C-G)—mediated by hydrogen bonds. Moreover, Watson and Crick postulated the sequence-specific arrangement of base pairs as the fundamental mechanism governing genetic information storage and transmission within the DNA molecule.

4. MECHANISMS OF DNA REPLICATION

In their seminal model of DNA structure and function, Watson and Crick hypothesized a mechanism of replication characterized by the unwinding of DNA strands facilitated by the disruption of hydrogen bonds between complementary base pairs—adenine with thymine, and cytosine with guanine. This process, termed **semiconservative replication**, generates two daughter DNA molecules, each harboring one original parental strand and one newly synthesized strand (Fig. **1A**). Dispersive replication does not necessarily require fragmentation. Subsequent experimental validation of this model by Meselson and Stahl corroborated the fidelity of semiconservative DNA replication, thereby elucidating the fundamental process underpinning genetic inheritance.

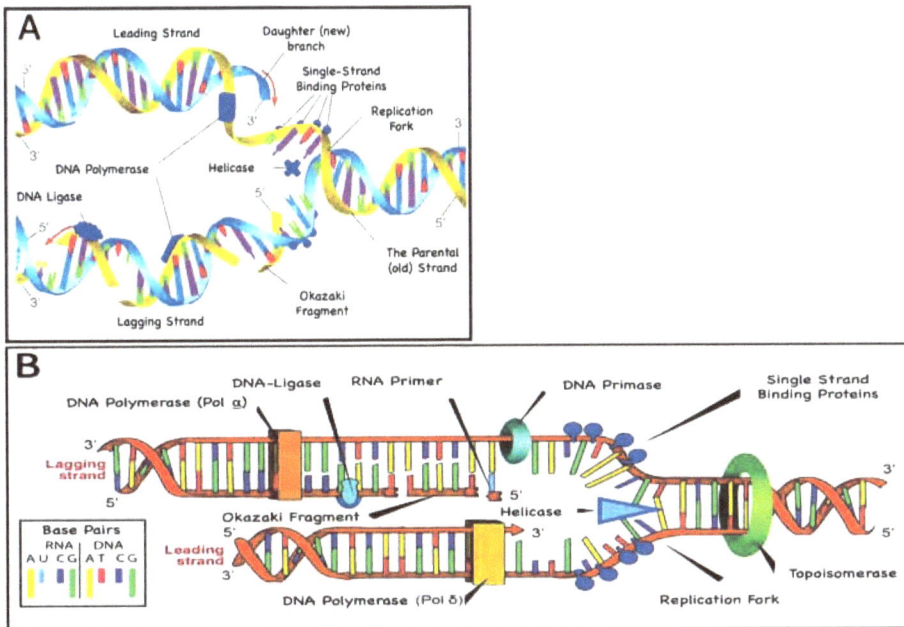

Fig. (1). DNA semiconservative replication and mechanisms of fork formation, (**A**) The semiconservative model of DNA replication proposed by Watson and Crick, depicting the unwinding of parental DNA strands, hydrogen bond disruption,and the formation of two daughter DNA molecules, each containing one parental and one newly synthesized strand. (**B**) DNA replication in Escherichia coli, illustrating the formation of replication forks within the circular chromosome, the unidirectional elongation of DNA strands, and the synthesis of the leading strand (continuous) and lagging strand (discontinuous, forming Okazaki fragments).

(**A**) The semiconservative model of DNA replication proposed by Watson and Crick, depicting the unwinding of parental DNA strands, hydrogen bond disruption, and the formation of two daughter DNA molecules, each containing one parental and one newly synthesized strand.

(**B**) DNA replication in Escherichia coli, illustrating the formation of replication forks within the circular chromosome, the unidirectional elongation of DNA strands, and the synthesis of the leading strand (continuous) and lagging strand (discontinuous, forming Okazaki fragments).

In *Escherichia coli*, the initiation of DNA replication heralds the formation of a distinctive bubble-like structure along the circular chromosome, facilitated by the assembly of **replication forks** (Fig. **1A**). Extensive investigations into bacterial DNA replication mechanisms have revealed a unidirectional elongation of the growing DNA strand exclusively in the 5' to 3' orientation, denoting the extension from the 5' carbon of one sugar moiety to the 3' carbon of the adjacent nucleotide. As DNA replication ensues, the leading strand undergoes continuous synthesis, while the lagging strand undergoes discontinuous synthesis, manifested in the formation of discrete segments termed **Okazaki fragments** (Fig. **1A** and **B**). This intricate process is orchestrated by **DNA polymerase** enzymes, which catalyze the sequential addition of nucleotides, aligning them along the template provided by the parental DNA strand (Fig. **1**).

Similarly, in eukaryotic cells, DNA replication adheres to analogous principles observed in prokaryotic systems. Spanning the vast expanse of elongated DNA molecules, replication proceeds bidirectionally from numerous replication origins, numbering in the hundreds or thousands. This coordinated replication from multiple initiation sites ensures the expeditious duplication of the entire genome while maintaining fidelity and accuracy in genetic inheritance.

Vocabulary (Index of Terms)

Adenine [/ˈædənɪn/]	Adenine is a nitrogenous purine base found in nucleic acids such as DNA and RNA, as well as in important coenzymes like NAD and FAD. It plays a crucial role in various cellular processes, including energy metabolism and genetic transcription.
Cytosine [/ˈsaɪtəsiːn/]	Cytosine is one of the four nitrogenous bases found in DNA and RNA, characterized by its pyrimidine structure. It pairs specifically with guanine through hydrogen bonding in DNA double helix. In RNA, cytosine pairs

with guanine as well. It plays a fundamental role in genetic coding and molecular recognition processes within living organisms.

DNA polymerase [/ˈdiːɛnˌeɪ ˈpɒlɪməreɪz/]

DNA polymerase is a crucial enzyme involved in DNA replication and repair processes. It catalyzes the formation of new DNA strands by adding complementary nucleotides to the template strand during replication. Additionally, DNA polymerase plays a pivotal role in proofreading and correcting errors that may occur during DNA synthesis, thereby ensuring the accuracy and fidelity of genetic information transmission.

Double helix [/ˈdʌbəl ˈhiːlɪks/]

The double helix refers to the characteristic three-dimensional structure of DNA, comprising two intertwined strands held together by hydrogen bonds between complementary nitrogenous bases. This iconic configuration facilitates the stable packaging of genetic information within the DNA molecule, ensuring its efficient storage and replication. Additionally, the double helix structure plays a crucial role in facilitating processes such as genetic transcription and molecular recognition within biological systems.

Guanine [/ˈgwɑːnɪn/]

Guanine is a nitrogenous base molecule characterized by its double-ring structure, found in both DNA and RNA. It forms complementary base pairs with cytosine, contributing to the stability and specificity of nucleic acid interactions. Guanine plays a pivotal role in genetic coding and molecular recognition processes within biological systems, facilitating essential cellular functions such as protein synthesis and genetic information transfer.

Nucleoside ['njuːkliəsaɪd]

A nucleoside is a compound composed of a nitrogenous base (either purine or pyrimidine) linked to a ribose or deoxyribose sugar molecule. Common nucleosides found in DNA and RNA include cytidine, cytosine deoxyriboside, thymidine, uridine, adenosine, adenine deoxyriboside, guanosine, and guanine deoxyriboside. It is worth noting that thymidine is a deoxyriboside, while cytidine, uridine, adenosine, and guanosine are ribosides. These molecules serve as the building blocks of nucleic acids and play crucial roles in genetic information storage and expression.

Okazaki fragment ['oʊkəzɑːki 'fræɡmənt]

An Okazaki fragment is a short, discontinuous segment of DNA synthesized during the replication of the lagging strand in DNA replication. These fragments are initiated by RNA primers and are subsequently synthesized by DNA polymerase. They are essential for the replication process, as they allow for the synthesis of both DNA strands in the 5' to 3' direction.

One-gene-one-enzyme hypothesis ['wʌn 'dʒiːn 'wʌn 'ɛnzaɪm haɪ'pɒθəsɪs]

The one-gene-one-enzyme hypothesis proposes that there exists a class of genes where each gene is responsible for the synthesis or regulation of a single enzyme. This hypothesis was formulated based on early genetic studies, but it has since been refined and replaced by the one-gene-one-polypeptide hypothesis. This revised hypothesis acknowledges that genes can also encode proteins other than enzymes, and that some proteins may be composed of multiple polypeptide chains, each encoded by a separate gene.

One-gene-one-polypeptide hypothesis [wʌn dʒin wʌn 'pɒlɪpaɪt ˌhaɪ'pɒθɪsɪs]

The one-gene-one-polypeptide hypothesis proposes that there is a broad category of genes where each individual gene regulates the synthesis of a single polypeptide chain. This polypeptide may function autonomously or as part of a larger, more complex protein structure. This hypothesis builds upon earlier theories, emphasizing the role of genes in determining the structure and function of proteins, recognizing that some proteins may consist of multiple polypeptide chains, each encoded by a distinct gene.

Purine ['pjʊəriːn]

Purine is a type of nitrogenous base characterized by a double-ring structure, serving as a fundamental component in the construction of nucleic acids and coenzymes. The two primary purines found in biological systems are adenine and guanine. Adenine and guanine are integral constituents of DNA and RNA molecules, participating in the encoding and transfer of genetic information. Additionally, they serve as essential components of various coenzymes involved in metabolic pathways and cellular processes.

Pyrimidine [paɪ'rɪmɪdiːn]

Pyrimidine is a heterocyclic organic compound with the chemical formula $C_4H_4N_2$, constituting the basic structure of pyrimidine bases. Several pyrimidine bases serve as essential components of nucleic acids. These bases play crucial roles in genetic material such as DNA and RNA, contributing to the encoding and transmission of genetic information.

Replication fork [ˌrɛplɪˈkeɪʃən fɔːrk]

A replication fork is a structure formed during DNA replication where the double helix of DNA is unwound and separated into two individual strands. At the replication fork, the DNA helicase enzyme catalyzes the unwinding of the DNA strands, creating two template strands for DNA synthesis. This process allows for the replication of both strands in a semiconservative manner, ensuring the accurate duplication of genetic information.

Semiconservative replication [ˌsɛmikənˈsɜːrvətɪv ˌrɛplɪˈkeɪʃən]

Semiconservative replication is a process of DNA replication where the DNA molecule separates into two strands, and each separated strand serves as a template for the synthesis of a new complementary strand. This method ensures that each newly formed DNA molecule retains one original ("parental") strand and one newly synthesized ("daughter") strand. This mechanism was elucidated by the classic Meselson-Stahl experiment and is fundamental to maintaining the fidelity of genetic information during cellular reproduction.

Thymine ['θaɪmiːn]

Thymine is a nitrogenous base molecule found in DNA but not in RNA. It is characterized by a single-ring structure and is complementary to adenine in the DNA double helix. Thymine forms hydrogen bonds specifically with adenine, contributing to the stability of the DNA molecule through base pairing. This base pairing specificity plays a crucial role in maintaining the integrity and fidelity of genetic information during DNA replication and transcription.

X-ray diffraction ['ɛksˌreɪ dɪˈfrækʃən]

X-ray diffraction is a scientific method used to investigate the arrangement of atoms within a crystalline substance. This technique involves directing a narrow beam of X-rays at the crystalline sample and analyzing the resulting diffraction patterns produced as the X-rays interact with the atomic structure of the material. X-ray diffraction has been instrumental in elucidating the three-dimensional structures of various biological molecules, including important macromolecules such as DNA, hemoglobin, and myoglobin. By interpreting the diffraction patterns, researchers can deduce valuable information about the spatial arrangement of atoms within these molecules, aiding in the understanding of their functions and properties.

COMPREHENSION TASKS

I. Key Terms: Matching

Instructions: Match each term from the left column with the corresponding description from the right column. Write the letter of the correct description next to the number of the term. Each description will only match with one term. Good luck!

1. 5' to 3' direction	**A.** A = T
2. Base	**B.** Unwinding must occur
3. Chargaff's rules	**C.** A DNA base
4. Cytosine	**D.** Ring structure composed of carbon and nitrogen
5. Double helix	**E.** The shape proposed by Watson and Crick
6. Nucleoside	**F.** A base plus a sugar
7. Okazaki fragment	**G.** Start of DNA replication
8. Replication fork	**H.** Small piece of DNA
9. Semiconservative replication	**I.** DNA chain lengthens
10. X-ray diffraction	**J.** Photographic process

Cell Biology True or False Quiz: Determine whether the following statements are true or false based on your understanding of cell biology concepts. Write "True" if the statement is correct and "False" if the statement is incorrect. Good luck!

1. ----- Hershey and Chase demonstrated that DNA constitutes genes.

2. ----- Complementation tests offer insights into metabolic pathways.

3. ----- Feulgen staining targets cellular structures.

4. ----- Robert Feulgen is credited with the discovery of nucleic acid.

5. ----- Phages consist solely of DNA and a protein envelope.

6. ----- Dispersive replication necessitates fragmentation.

7. ----- Semiconservative replication isn't universal across all organisms.

8. ----- Auxotrophic strains are observed in mutant Neurospora.

9. ----- Virulent pneumonia bacteria produce a protective capsule.

10. ----- Watson and Crick's model of DNA structure did not rely on X-ray diffraction data.

11. ----- Genetic information is encoded in the sequence of bases along the DNA molecule.

II. **Completion Exercise: Complete each sentence by filling in the blanks with the appropriate term or phrase**. Write your answers in the space provided. Good luck!

1. Beadle and Tatum were the first to demonstrate a correlation between _____ and _____.

2. Linus Pauling formulated his one-gene-one-polypeptide hypothesis utilizing an electrical molecular separation technique known as _____.

3. The identification of DNA's composition of four bases stemmed from the investigations of _____.

4. The combination of a base and a sugar constitutes a _____.

5. The fundamental characteristics of DNA were outlined in a series of principles postulated by _____.

6. DNA's ability to self-replicate relies on each strand's capacity to function as a _____ when separated.

7. The model of DNA replication proposed by Watson and Crick, which involves the unwinding of the double helix, is referred to as the _____ model.

8. Okazaki fragments are linked together by the enzyme _____.

III. Multiple Choice: Choose the best answer for each question by circling the corresponding letter.

1. The discovery of genes' role in cellular metabolism was pioneered by _____.

a. Garrod

b. Watson

c. Tatum

d. Ingram

e. None of the above

2. Electrophoresis is a technique used to _____.

a. Separatethe nuclear material of a cell from the remainder of the cell

b. Allow scientists to stain molecules

c. Separate molecules on the basis of their electric charge

d. Allow scientists to determine the spatial arrangements of atoms in molecules

e. Allow scientists to determine the lengths of chemical bonds

3. Pauling's application of electrophoresis in sickle cell anemia research led to the conclusion that _____.

a. Polypeptides carry the genetic information

b. The sickle cell gene chemically alters the hemoglobin protein

c. DNA carries the genetic information

d. The one-gene-one-enzyme hypothesis was incorrect

e. None of the above.

4. Feulgen staining induces a color change in nucleic acids, turning them into _____.

a. Deep red

b. Pink

c. Deep green

d. Black

e. Brown

5. Pauling's research prompted the revision of the one-gene-one-enzyme hypothesis to the currently embraced _____ .

a. One-gene-one-protein hypothesis

b. One-gene-one-polypeptide hypothesis

c. One-gene-one-amino-acid hypothesis

d. One-gene-one-alpha-chain hypothesis

e. None of the above

6. The revelation in the early 1900s that the nucleus contains nucleic acids and proteins sparked a widespread belief that _____ .

a. Genes are in the nucleus

b. Genes are broken down by pepsin

c. Genes are made of nucleic acid

d. Genes are made of protein

e. None of the above

7. Disease-causing strains of microorganisms are characterized by _____ .

a. Nonvirulent

b. Lethal

c. Virulent

d. Disgenic

e. None of the above

8. Griffith's experiments with pneumococci revealed that the difference between the S and R strains lay in _____ .

a. The R strain never became virulent

b. The S strain formed rough harmless colonies

c. A polysaccharide capsule was present on the S strain

d. The S strain was harmless

e. All of the above

9. Chase and Hershey's experiments in the 1950s demonstrated that _____.

a. Proteins carry the genetic information

b. DNA carries the genetic information

c. Proteins control DNA replication

d. Polypeptides are the units of inheritance

e. Enzymes carry genetic information

10. Nucleotides serve as the fundamental units of the DNA molecule. Each nucleotide comprises _____.

a. A double-stranded molecule

b. Four nitrogenous bases

c. A nitrogenous base, a phosphate group, and a sugar molecule

d. Four sugar molecules

e. Four phosphate groups

11. The four nucleotides constituting the DNA molecule differ in that _____.

a. Each base contains a different sugar

b. Each nucleotide contains a different phosphate group

c. There are four different sugars in the molecule

d. Each nucleotide contains a different nitrogenous base

e. Each nucleotide contains a different sugar group

12. Within the DNA molecule, _____.

a. Adenine and guanine are purines

b. Cytosine and thymine are pyrimidines

c. Adenine bonds to thymine

d. Guanine bonds to cytosine

e. All of the above

13. Chargaff's rules articulate that _____.

a. In the DNA molecule, the number of adenines = number of thymines and the number of guanines = number of cytosines

b. In the DNA molecule, the number of gua-nines and cytosines = the number of adenines and thymines

c. The ratio of $(A+T)/(G+C)$ is constant for a species

d. All of the above

e. a and c only

14. The double helix model of DNA structure was proposed by _____.

a. Stahl

b. Watson and Crick

c. Beadle and Tatum

d. Avery and McLeod

e. Pauling

15. All of the following contributed to Watson and Crick's elucidation of DNA structure except _____.

a. X-ray diffraction

b. Cardboarµ model building

c. Levene's ideas on DNA components

d. Pauling's work on DNA structure

e. Cesium chloride density gradients

16. The phosphate group in the DNA molecule is _____.

a. Links two nucleotide bases together

b. Links the 5-C sugar of one nucleotide to the 3-C sugar of the next nucleotide

c. Links two strands of the DNA molecule to-gether

d. Links two nucleotide bases together

e. None of the above

17. Nucleotide bases on one strand of the DNA molecule are connected to bases on the second strand through_____.

a. Covalent bonds

b. Weak hydrogen bonds

c. Ionic bonds

d. Phosphate groups

e. None of the above

18. DNA replication is described as semiconservative because_____.

a. Only the bases are replicated

b. The bases are replicated in small fragments

c. Only half the DNA molecule is replicated at any one time

d. Each new molecule has one strand from the original molecule

e. The two stands of the molecule are replicated

19. The strands of the DNA molecule are antiparallel, meaning that_____.

a. Both strands are oriented in a 5' to 3' direction

b. Both strands are oriented in a 3' to 5' direction

c. The molecule is oriented differently from the rest of the cell

d. Only one strand of the molecule replicates

e. One strand is oriented in a 5' to 3' direction and the other in a 3' to 5' direction

20. DNA chains extend in the _____ direction.

a. 3' to 5' direction

b. 5' to 3' direction

c. 3' direction

d. 5' direction

e. None of the above

READING UNDERSTANDING

Atmospheric pressure can uphold a water column of approximately 10 meters, yet plants exhibit a remarkable ability to transport water to greater heights. Notably, sequoia trees can propel water upwards, surpassing 100 meters above ground level. The mystery surrounding water movement in trees and tall plants persisted until the late 19th century. Initially, botanists speculated that living plant cells acted as pumps, but numerous experiments debunked this notion by showcasing water movement in plants devoid of living cells. Alternative explanations, such as root pressure exerted from the plant's base, failed to account for the ascent of water in tall trees, particularly conifers with notably low root pressure.

In pondering how water ascends tall trees without being pumped or pushed, the cohesion-tension theory emerges as the prevailing explanation. According to this theory, water is drawn upward through the plant. This pull originates from the evaporation of water at the plant's apex, generating negative pressure or tension. As water evaporates from leaf surfaces, it creates a vacuum, prompting water from within the plant to rise in continuous columns from roots to canopy. These columns are upheld by cohesive forces between water molecules, akin to surface tension. In narrow tubes, water's cohesive strength is so potent that it rivals that of a steel wire of equivalent diameter, enabling water columns to ascend to considerable heights without fracturing.

1. How many theories are mentioned by the author?

A. One

B. Two

C. Three

D. Four

2. Which question is addressed by the passage?

A. What is the effect of atmospheric pressure on foliage?

B. When do dead cells harm plant growth?

C. How does water get to the tops of trees?

D. Why is root pressure weak?

3. The term "demonstrated" in the first paragraph is closest in meaning to ___.

A. Ignored

B. Showed

C. Disguised

D. Distinguished

4. How do botanists confirm that root pressure is not the sole force moving water in plants?

A. Some very tall trees have weak root pressure

B. Root pressures decrease in winter

C. Plants can live after their roots die

D. Water in a plant's roots is not connected to water in its stem

5. Which statement is supported by the passage?

A. Water is pushed to the tops of trees

B. Botanists have proven that living cells act as pumps

C. Atmospheric pressure draws water to the tops of tall trees

D. Botanists have changed their theories of how water moves in plants

6. What triggers the tension that draws water up a plant?

Plant growth

Root pressure

Evaporation

Photosynthesis

7. According to the passage, why does water travel through plants in unbroken columns?

A. Root pressure moves the water very rapidly

B. The attraction between water molecules is strong

C. The living cells of plants push the water molecules together

D. Atmospheric pressure supports the columns

Unraveling the Precision of DNA Replication

1. Exploring the Mystery of DNA's Accuracy

As the pivotal role of DNA as the blueprint of life became evident, scientists were faced with the enigmatic puzzle of how such a seemingly basic molecule could orchestrate the intricacies of biological functions. DNA, at its core, consists of a repetitive sequence of identical five-carbon sugars intricately linked together, each sugar hosting one of four organic bases: adenine (A), guanine (G), thymine (T), and cytosine (C).

Initially perceived as a monotonous sequence akin to AGTC AGTC AGTC, the complexity of DNA's involvement in heredity baffled early researchers. However, by the late 1940s, meticulous analyses conducted by Erwin Chargaff and his team at Columbia University unveiled a complex reality: the composition of DNA varied significantly among different organisms, challenging the notion of its simplicity. Chargaff's observations revealed a fundamental consistency: the quantity of adenine always equaled that of thymine, and likewise, the quantity of guanine mirrored that of cytosine across DNA molecules. However, the combined quantity of adenine and thymine often diverged markedly from that of guanine and cytosine.

These revelations by Chargaff and others catalyzed a paradigm shift in our understanding of genetic inheritance, underscoring the intricate interplay between DNA's molecular components and its pivotal role in preserving life's diversity.

Upon learning of Franklin's results in 1953, James Watson and Francis Crick, young investigators at Cambridge University, swiftly unraveled the probable structure of DNA. Employing deductive reasoning, they meticulously constructed models of nucleotides and explored how these units could assemble into a molecule consistent with DNA's known structure. Their breakthrough came with the proposal of a simple double helix, where hydrogen bonds between complementary bases stabilized the structure. The Watson-Crick model elucidated the complementary nature of DNA replication, where each strand serves as a template for synthesizing a new complementary strand during cell division.

The predictive power of the Watson-Crick model was confirmed in 1958 by Matthew Meselson and Frank Stahl of the California Institute of Technology. Their seminal experiment, involving bacterial growth in heavy nitrogen followed by transfer to normal nitrogen, demonstrated the semiconservative nature of DNA replication. Analysis revealed that after replication, each daughter DNA duplex contained one heavy and one light strand, confirming the model's prediction and solidifying our understanding of DNA replication.

2. *Unveiling the Identity of the Genetic Code*

As the scientific community recognized DNA's pivotal role as the carrier of hereditary information, the quest to decipher its contents intensified. Since Mendel's pioneering work with pea traits, geneticists grappled with understanding the nature of the information stored within DNA. The traits observed by Mendel, such as pea colors and shapes, represented the end result of intricate processes. But what alterations in the hereditary information would lead to changes in these Mendelian traits?

The first clues emerged shortly after Mendel's experiments, though their significance was not immediately apparent. In 1902, British physician Archibald Garrod, collaborating with geneticist William Bateson, noted the prevalence of certain diseases within specific families. Upon examining multiple generations, they observed that some disorders followed a Mendelian inheritance pattern, controlled by simple recessive alleles. Garrod inferred that these disorders were hereditary traits resulting from past alterations in the genetic information, passed down through generations.

Further investigations by Garrod into disorders like alkaptonuria revealed profound insights. Patients with alkaptonuria excreted urine that rapidly darkened upon exposure to air due to the presence of homogentisic acid. Garrod discerned that these patients lacked the enzyme necessary to break down homogentisic acid, suggesting that inherited disorders might stem from enzyme deficiencies.

The logical leap from Garrod's findings to the notion that DNA encodes enzymes was further substantiated in 1941 by Stanford University geneticists George Beadle and Edward Tatum. Their pioneering experiments with the bread mold *Neurospora* provided definitive evidence. By inducing mutations in the mold's chromosomes and studying their effects, Beadle and Tatum demonstrated a clear relationship between genes and enzymes.

The choice of *Neurospora* as an experimental organism facilitated the clarity of Beadle and Tatum's results. By cultivating various strains of the fungus on defined media, they identified mutants deficient in specific metabolic pathways. By adding various chemicals to the minimal medium, they pinpointed the nature of these deficiencies, revealing that each mutant harbored a defect in a particular enzyme.

These experiments led Beadle and Tatum to propose the "one gene-one enzyme" hypothesis, asserting that genes dictate the structure of enzymes. Subsequent research, including Frederick Sanger's sequencing of insulin amino acids in 1953, and Vernon Ingram's analysis of sickle cell anemia in 1956, further affirmed this notion. Ingram's work elucidated that a single amino acid change in hemoglobin led to the manifestation of sickle cell anemia, providing concrete evidence of how genetic mutations can affect protein structure and function.

These breakthroughs collectively clarified the unit of hereditary information encoded within DNA: the sequence of nucleotides specifying the amino acid sequence of proteins or polypeptides. This sequence, known as a gene, forms the basis of genetic inheritance and serves as the blueprint for the synthesis of essential biological molecules.

Section 3

The Diversity of Life - Development, Ecology, and Evolution

Biology for Students, 2025, 89-115

CHAPTER 5

Life's Story: From Start to Diversity

Abstract: The origins of life on Earth are believed to have occurred after the planet's formation, around 4.6 billion years ago. Life itself is thought to have emerged approximately 3.8 to 4 billion years ago, following a period of chemical evolution that led to the formation of organic molecules and proto-cells. This transition to life is marked by the development of complex molecular structures, including the role of RNA as a likely precursor to DNA in early life forms. As Earth's conditions became more conducive to life, cellular organisms evolved, starting with prokaryotic life and later giving rise to eukaryotes. Over time, life diversified into distinct domains, with advances in molecular biology leading to the adoption of the three-domain system—Bacteria, Archaea, and Eukarya—replacing the older five-kingdom classification system. This modern framework reflects our current understanding of the evolutionary relationships among life forms. The development of taxonomy, with its hierarchical structure, continues to evolve as we refine our understanding of the interconnectedness and evolutionary history of all living organisms.

Keywords: Chemical evolution, Eukaryotic cells, Phylogenetic trees, Proto-cells, Taxonomy.

UNDERSTANDING THE CHAPTER'S FOCUS: A ROADMAP

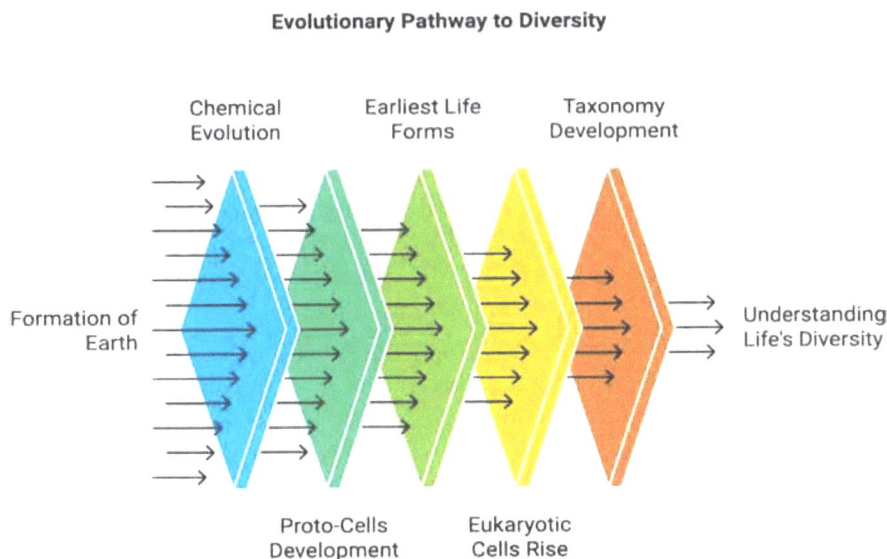

Evolutionary Pathway to Diversity

Mohammad Mehdi Ommati

1. HOME FOR LIFE: THE BIRTH OF OUR SOLAR SYSTEM AND PLANET EARTH

The story of life's origins begins after the formation of Earth, approximately 4.6 billion years ago, when conditions became suitable for life to emerge. This narrative starts with a colossal event known as the **Big Bang**, marking the inception of our universe around 13.8 billion years ago. About 5 billion years ago, the sun, the centerpiece of our solar system, began to form from a vast cloud of primordial matter. Following this, planets like Earth began to emerge through a process known as accretion, around 4.6 billion years ago. Earth itself is characterized by distinct layers, comprising a solid **crust**, a semi-solid **mantle**, and a predominantly molten **core** with a solid inner region. Crucial aspects such as Earth's size, temperature, composition, and orbital distance from the sun played pivotal roles in shaping its suitability for life.

Scientists propose that life originated on early Earth through a process known as chemical evolution, where simple non-living molecules gradually became more complex, eventually giving rise to the first forms of life. This likely occurred around 3.8 to 4 billion years ago, after billions of years of chemical processes, thus ushering in the dawn of biological complexity.

2. THE RISE OF LIFE: FROM CHEMICAL BUILDING BLOCKS TO PROTO-CELLS

The journey towards life's emergence is marked by the intricate dance of organic and biological molecules on the primordial stage of our planet. Laboratory experiments, inspired by the conditions of early Earth, have yielded valuable insights into the pre-life stages of chemical organization. Pioneering endeavors, notably the experiments of Miller and Urey, have demonstrated the formation of organic monomers like amino acids, simple sugars, and nucleic acid bases under simulated early Earth conditions. These fundamental building blocks set the stage for the subsequent assembly of larger polymers such as **proteinoids** and nucleic acids, potentially on clay or rock surfaces.

Studies have unveiled three primary types of organic molecular aggregates that could have played pivotal roles in the transition from non-life to life. Aleksandr Oparin's research yielded polymer-rich droplets known as **coacervates**, and Sidney Fox's experiments produced proteinoid microspheres from amino acids and water mixtures, and **liposomes**, spherical lipid bilayers, formed from phospholipids. It is hypothesized that one or more of these aggregates may have served as precursors to the earliest cellular structures.

As the narrative of life's emergence unfolds, attention turns to the development of RNA and DNA as indispensable informational molecules. RNA, with its ability to spontaneously form under conditions akin to those of early Earth, is postulated to have been the initial bearer of biological information. The discovery of RNA ribozymes, molecules capable of catalytic activity akin to enzymes, suggests a potential mechanism for assembling new RNAs from early nucleotides. These catalytic RNAs might have also facilitated crucial processes like RNA-mediated exchanges, akin to primitive forms of genetic recombination.

Beyond the realm of informational molecules, the journey towards cellular life involves the formation of lipid-protein surface layers, the establishment of the genetic code, the sequestering of RNA or DNA within cell-like structures, and the development of metabolic pathways. Each of these milestones represents a crucial step in the complex tapestry of life's evolution from simple molecular aggregates to the dawn of cellular complexity.

3. EARLY LIFE FORMS: TRACING THE ORIGIN OF CELLS

The quest to uncover the earliest traces of cellular life leads us to rocks dating back approximately 3.5 billion years, where we find fossils that may hold the secrets of life's humble beginnings. These ancient relics suggest that anaerobic heterotrophs were the first cells to populate our planet, with autotrophs emerging later and playing a critical role in shaping Earth's atmosphere. These pioneering autotrophs, capable of self-nourishment, played a pivotal role in shaping Earth's atmosphere by releasing oxygen as a metabolic by-product. This oxygenation led to the formation of the **ozone layer**, a protective shield against harmful ultraviolet radiation, allowing life to flourish in shallow waters and on land surfaces.

The rise in atmospheric oxygen levels also catalyzed the evolution of aerobic cells and ushered in the era of cellular respiration, marking the onset of the global carbon cycle. As oxygen levels increased, aerobic cells thrived, paving the way for diverse metabolic pathways and ecological niches to emerge. This transformative phase in Earth's history laid the groundwork for the intricate web of life that we observe today.

While the earliest cells were exclusively prokaryotic in nature, approximately 1.5 billion years ago, a significant evolutionary leap occurred with the appearance of eukaryotic cells. These eukaryotes, distinguished by their complex internal structures and membrane-bound organelles, heralded a new chapter in the saga of life's diversity and complexity.

4. EARTH'S DYNAMIC EVOLUTION: SHAPING LIFE'S JOURNEY

The evolutionary tapestry of life on Earth is intricately woven with the threads of geological transformations, shifts in oceanic landscapes, and fluctuations in climate. At the heart of our planet lies a complex structure, comprising a relatively thin, solid crust enveloping a semi-fluid mantle, which in turn encases a partially molten core. This dynamic arrangement sets the stage for a phenomenon known as **continental drift**, where vast segments of the Earth's crust, akin to gigantic jigsaw pieces, glide over the semi-solid mantle.

Over the course of the last 500 million years, this relentless movement of tectonic plates has reshaped the face of our planet, giving rise to the diverse array of continents that adorn our world map today. However, this geological ballet is not a silent one; it orchestrates symphonies of change that reverberate throughout the biosphere. As continents collide and rift apart, climatic conditions undergo dramatic shifts, triggering waves of mass extinctions and reshaping the destinies of countless organisms.

These geological dramas are further underscored by the dance of the celestial bodies. Variations in Earth's orbit and fluctuations in solar energy output have cast shadows of glaciation across the planet, punctuating the epochs of life with periods of icy embrace. In this ever-changing theater of existence, organisms must adapt or perish, navigating the tumultuous currents of geological and climatic change to ensure their survival.

5. UNVEILING NATURE'S DIVERSITY: THE SCIENCE OF TAXONOMY

At the heart of biology lies taxonomy, the systematic effort to organize and classify the vast array of life forms that populate our planet. Building on the foundation laid by Linnaeus, biologists employ the **binomial system of nomenclature** to assign each organism a unique identity, consisting of a **genus** and **species** designation. This hierarchical system extends beyond individual species, encompassing broader taxonomic categories such as **domain, kingdom, phylum** (in animals), **division** (in plants), **class, order, family, genus**, and **species** (Fig. **1**).

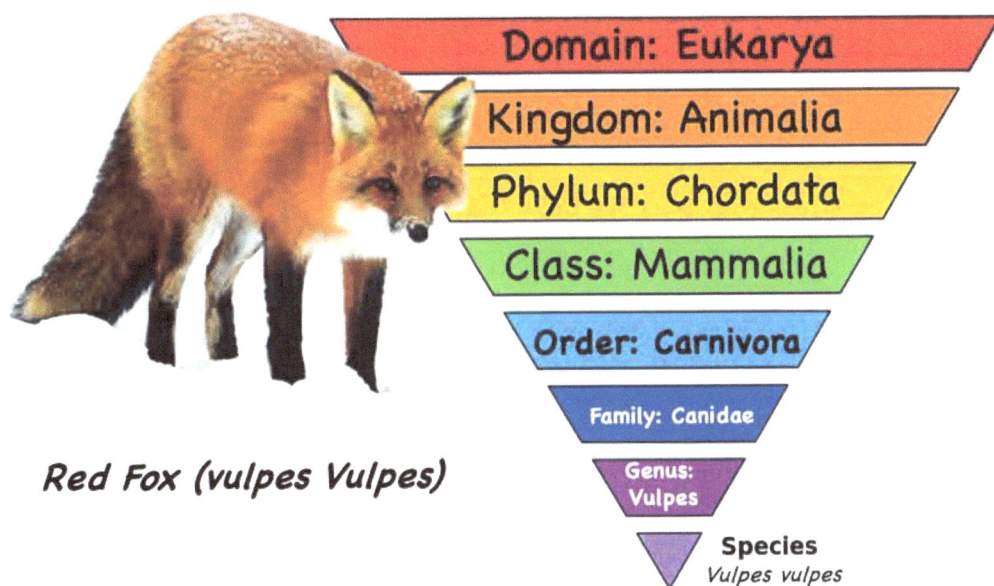

Fig. (1). Taxonomic Rank of Red Fox (*Vulpes vulpes*).

Taxonomic classification draws upon a wealth of evidence from various biological disciplines, including biochemistry and comparative anatomy, to delineate species and higher **taxa (taxon)**. While early taxonomic frameworks relied heavily on morphological traits, modern biology embraces the criterion of reproductive isolation to define species, reflecting a deeper understanding of evolutionary relationships.

Indeed, **taxonomy** serves as a window into the evolutionary tapestry of life. **Clades**, defined as taxonomic units comprising organisms descended from a common ancestor, offer insights into the interconnectedness of living beings and their shared evolutionary heritage. Through the lens of taxonomy, we gain a deeper appreciation for the diversity of life forms and the intricate relationships that bind them across the tree of life.

6. MAPPING EVOLUTIONARY LINEAGES: THE TRANSITION FROM FIVE KINGDOMS TO THREE DOMAINS

In the pursuit of understanding life's evolutionary history, biologists employ phylogenetic trees as visual frameworks to depict the interconnectedness of organisms and their evolutionary relationships. The classification of living organisms has evolved significantly with advances in molecular biology. Historically, the five-kingdom system was widely used, categorizing life forms into distinct kingdoms: Monera, Protista, Fungi, Plantae, and Animalia. However, with the advent of molecular techniques and a better understanding of genetic relationships, biologists now favor the three-domain system, which divides life into three broad categories: Bacteria, Archaea, and Eukarya. These domains are based on fundamental differences in cellular structure and genetic makeup, and they reflect a more accurate understanding of the evolutionary history of life on Earth.

While this classification system serves as a practical tool for organizing the diversity of life, it is important to note that the five kingdoms may not represent true clades-groups of organisms descended from a common ancestor. Instead, they offer a simplified framework for categorizing organisms based on shared characteristics and evolutionary traits. As our understanding of evolutionary relationships continues to evolve, taxonomic frameworks may undergo refinement, reflecting new insights into the complex tapestry of life's diversity.

CONCLUSION WITH EXAMPLES IN COW AND MOUSE

In the grand tapestry of life's taxonomy, organisms are meticulously categorized into a structured hierarchy that reflects their evolutionary relationships and distinctive traits. At the highest echelon, we encounter the three domains: Bacteria, Archaea, and Eukarya. Within Eukarya, both cows (*Bos taurus*) and mice (*Mus musculus*) find their place, exemplifying the vast diversity encapsulated within this domain. Transitioning down the ranks, we arrive at the Kingdom Animalia, a realm that unites these creatures under the common banner of multicellularity and heterotrophy. The Phylum Chordata then beckons, encompassing animals with a notochord at some developmental stage, a defining feature shared by both bovines and rodents. As we delve deeper into the classification system, the Class Mammalia warmly embraces cows and mice, recognizing their endothermic nature and milk-nursing habits. Within the Order Artiodactyla, cows graze peacefully, while mice scurry within the Order Rodentia, illustrating the diverse evolutionary paths these creatures have traversed. In the familial embrace of Bovidae, cows find their place, while mice seek kinship within the family Muridae. The genus *Bos cradles* cows within its fold, while mice find their identity within the genus Mus. Finally, at the species level, cows are distinguished as *Bos taurus*, and mice as *Mus musculus*, underscoring the intricate and nuanced classification system that illuminates the interconnectedness and diversity of life forms on our planet. An example of the cow classification is provided in Table **1**.

Table 1. Example for Cow.

1. Domain	→	**Eukarya**
2. Kingdom	→	**Kingdom Animalia**
3. Phylum	→	**Cordata**
4. Class	→	**Mammalia**
5. Order	→	**Artiodactyla**
6. Family	→	**Bovidae**
7. Genus	→	**Genus Bos**
8. Species	→	***Bos taurus***

Vocabulary (Index of Terms)

Big Bang [bɪg bæŋ]

The Big Bang refers to the prevailing cosmological theory explaining the origin and evolution of the universe. It posits that the universe began as an extremely hot and dense point roughly 13.8 billion years ago, then rapidly expanded and cooled, giving rise to the formation of galaxies, stars, and other celestial structures over time. This theory suggests that the universe continues to expand to this day.

Binomial system of nomenclature [baɪ'noʊmiəl 'sɪstəm əv noʊmən'kleɪtʃər]

The binomial system of nomenclature is a method used in biology to assign scientific names to organisms, consisting of two parts: the genus name and the species epithet. This system, developed by Carl Linnaeus, provides a standardized way of naming and classifying living organisms, allowing for clear communication and understanding among scientists worldwide.

Clade [kleɪd]

In cladistics, a clade refers to a group of organisms that include an ancestor and all of its descendants. Clades are identified based on shared characteristics, called synapomorphies, which are traits inherited from a common ancestor. This method of classification focuses on evolutionary relationships, grouping organisms based on their evolutionary history rather than on traditional morphological similarities.

Class [klæs]

In biological classification, a class is a major taxonomic rank used to categorize living organisms based on shared characteristics. It represents a group that contains several similar or closely related orders, although sometimes it may consist of only one order. Classes are typically large and distinct groups within the hierarchical classification system, facilitating the organization and understanding of the diversity of life forms.

Coacervate [ˌkoʊ.əˈsɜr.veɪt]

Coacervates are spherical structures formed by the aggregation of organic macromolecules within a solution, where water molecules align around them. These macromolecules, often polymers such as proteins or nucleic acids, undergo phase separation in aqueous environments, leading to the formation of distinct coacervate droplets. These droplets are surrounded by a layer of water molecules, creating a boundary between the coacervate and its surrounding environment. Coacervates are of interest in the study of the origin of life, as they mimic some properties of primitive cells and may have played a role in the early stages of biochemical evolution.

Continental drift [ˌkɒn.tɪˈnɛn.təl drɪft]

Continental drift is the scientific theory proposing that the Earth's continents were once part of a single supercontinent, which later fragmented and drifted apart to their current positions. According to this theory, the movement of the Earth's tectonic plates over geological time scales has led to the gradual shifting of continents across the Earth's surface. This process has resulted in the formation of the present-day continents and the arrangement of landmasses as observed today. The theory of continental drift revolutionized geology and provided key insights into the geological processes shaping the Earth's surface.

Core [kɔːr]

The core of the Earth refers to the innermost layer of the planet, situated beneath the mantle. It is composed primarily of iron and nickel, with a solid inner core surrounded by a liquid outer core. The core plays a crucial role in generating the Earth's magnetic field through the process of convection and the movement of molten metals.

Crust [krʌst]

The crust of the Earth is the outermost layer of the planet, characterized by its solid composition and varying thickness. It is divided into two main types: continental crust, which forms the continents and is thicker but less dense, and oceanic crust, which underlies the ocean basins and is thinner but denser. The crust is crucial for supporting life as it provides a habitat for terrestrial organisms and serves as the surface upon which geological processes such as erosion, weathering, and sedimentation occur.

Division [dɪ'vɪʒ(ə)n]

In the classification of plants, a division is a taxonomic unit that groups together several related classes or, in some cases, comprises only one class with distinct shared characteristics. Divisions are used to categorize plants based on key morphological and anatomical features, as well as reproductive structures. Each division represents a significant level of evolutionary divergence within the plant kingdom, reflecting the diversity and complexity of plant life.

Family ['fæmɪli]

In the classification of both plants and animals, a family is a taxonomic grouping that encompasses several closely related genera or, occasionally, comprises only one genus with similar characteristics. Family names typically end in -aceae or -ae in botany and in -idae in zoology. Families represent an intermediate level of classification between the genus and order, providing insights into the evolutionary relationships and shared ancestry among organisms within a broader taxonomic context.

Genus ['dʒiːnəs]

(*plural, genera*) In the classification of both plants and animals, a genus is a taxonomic category that includes several closely related species sharing common characteristics. Members of the same genus often exhibit distinct similarities in their morphology, behavior, and genetic makeup. The genus serves as an intermediate level of classification between family and species, providing a framework for organizing and understanding the diversity of life forms within a broader taxonomic context.

Kingdom ['kɪŋdəm]

In biology, a kingdom is one of the highest taxonomic categories used to classify living organisms, representing major divisions based on fundamental differences in cellular structure, mode of nutrition, and reproductive strategies. While the traditional five-kingdom system

(Animalia, Plantae, Fungi, Protista, and Monera) was once standard, it has since been largely replaced by the three-domain system, which groups organisms into Bacteria, Archaea, and Eukarya. This more modern framework reflects deeper insights from molecular biology, offering a more precise understanding of the evolutionary relationships among different forms of life.

Liposome ['lɪpəsəʊm]

A liposome is a spherical vesicle formed by the self-assembly of phospholipids in aqueous solutions. These bilayered structures mimic the composition and properties of biological membranes and serve as models for studying cellular processes such as drug delivery, gene therapy, and membrane fusion. Liposomes can encapsulate hydrophilic or hydrophobic molecules within their lipid bilayers, making them versatile vehicles for delivering therapeutic agents and other bioactive compounds.

Mantle ['mæntl]

The mantle is the region of the Earth's interior located between the crust and the core. Comprising a thick layer of solid rock, the mantle plays a crucial role in geological processes such as plate tectonics, convection currents, and mantle plumes. It is divided into two main regions: the upper mantle, which is solid but capable of flowing over long periods of time, and the lower mantle, which is more rigid and extends to the boundary with the Earth's liquid outer core. The mantle's dynamic behavior influences the Earth's surface features, seismic activity, and heat distribution.

Order ['ɔːdər]

In the classification of both plants and animals, an order is a taxonomic unit that groups together several similar or closely related families, or in some cases, comprises only one family with distinct shared characteristics. Orders represent a higher level of organization within the hierarchical classification system, offering insights into the evolutionary relationships and functional traits of organisms. The names of orders typically end in -ales for plants and -a for animals, reflecting their systematic placement within the broader taxonomic framework.

Ozone layer ['oʊzoʊn 'leɪər]

The ozone layer is a region of the Earth's atmosphere located approximately 20 to 50 kilometers above the surface, characterized by the high concentration of ozone molecules. Ozone in this layer is primarily formed through the interaction of ultraviolet (UV) radiation from the sun with oxygen molecules. The ozone layer plays a crucial role in shielding the Earth from harmful UV radiation by absorbing and filtering out much of the incoming solar radiation. This protective function helps to safeguard living organisms on the Earth's surface from the damaging effects of UV rays, including skin cancer, cataracts, and disruption of ecosystems.

Phylum ['faɪləm] (pl. phyla ['faɪlə])

In animal classification, a phylum is a major taxonomic category that groups together several classes sharing common structural and developmental characteristics. Phyla represent a higher level of organization within the hierarchical classification system, reflecting broad patterns of evolutionary relationships among diverse animal groups. Each phylum encompasses a diverse array of organisms descended from a common ancestor, with members exhibiting variations in morphology, physiology, and ecological adaptations. The classification of animals into phyla provides a framework for understanding the evolutionary history and diversity of life forms in the animal kingdom.

Proteinoid ['proʊtiːnɔɪd]

A proteinoid is a protein-like structure composed of branched chains of amino acids, forming the basic framework of a microsphere. These amino acid chains undergo polymerization reactions under certain conditions, leading to the formation of proteinoids. Proteinoids exhibit properties similar to those of proteins but lack the complex folding patterns and specific sequences characteristic of true proteins. They are often synthesized under conditions simulating primitive Earth environments, suggesting a possible role in the origin of life. Proteinoids are associated with the formation of protocells and the emergence of early cellular structures, contributing to theories on the origin of life on Earth.

Species ['spiːʃiːz]

In the classification of both plants and animals, a species is a fundamental taxonomic unit comprising a group of organisms capable of interbreeding and producing fertile offspring under natural conditions. Species are defined by their ability to exchange genetic material and maintain reproductive compatibility within their populations. Each species exhibits distinct morphological, physiological, and behavioral traits that distinguish it from other closely related species. The concept of species plays a central role in evolutionary biology, serving as the primary unit of diversity and adaptation over time.

Taxon ['tæksɒn]

A taxon refers to any rank or category used in the hierarchical classification of organisms, encompassing groups ranging from species to kingdoms. Taxa are organized based on shared similarities and evolutionary relationships among organisms, providing a systematic framework for organizing biodiversity. Each taxon represents a distinct level of classification within the hierarchical scale, facilitating the categorization and study of biological diversity across various levels of organization.

Taxonomy [tæk'sɒnəmi]

Taxonomy is the scientific study of the principles, methods, and rules governing the classification and categorization of organisms according to their similarities and differences. It encompasses the identification, naming, and organization of living organisms into hierarchical groups based on shared characteristics and evolutionary relationships. Taxonomy serves as a fundamental discipline in biology, providing a systematic framework for understanding the diversity of life on Earth and elucidating the evolutionary history and relationships among different species and higher taxonomic groups.

COMPREHENSION TASKS

I.Key Terms: Matching

Instructions: Match each term from the left column with the corresponding description from the right column. Write the letter of the correct description next to the number of the term. Each description will only match with one term. Good luck!

1.	Stromatolite	**A.**	Composed of iron and nickel
2.	Coacervate	**B.**	The middle layer of earth
3.	Clade	**C.**	Fossil marine algae
4.	Ozone layer	**D.**	Outer rock skin
5.	Pangaea	**E.**	Drifting of the crust through activation in the mantle
6.	Crust	**F.**	First binomial name
7.	Species	**G.**	Monophyletic group
8.	Core	**H.**	Branching of closely related groups
9.	Plate tectonics	**I.**	Share a common ancestor.
10.	Cladistics	**J.**	Linnaeus
11.	Big bang	**K.**	Single land mass
12.	Monophyletic	**L.**	Filers out ultraviolet light
13.	Binomial system	**M.**	Polymer-rich spheres
14.	Mantle	**N.**	Origin of universe

Cell Biology True or False Quiz: Determine whether the following statements are true or false based on your understanding of cell biology concepts. Write "True" if the statement is correct and "False" if the statement is incorrect. Good luck!

1. ----- The formation of the Earth and the Sun occurred shortly after the occurrence of the Big Bang.

2.   ----- The Earth's center is primarily comprised of molten iron and nickel.

3.   ----- Water vapor in the atmosphere is responsible for the greenhouse effect.

4.   ----- The hypothesis of special creation cannot be tested.

5.   ----- It is impossible to create lipid bilayers in laboratory settings.

6.   ----- It is likely that the initial living cells were aerobic heterotrophs.

7.   ----- The emergence of the first cells may have been dependent on the shielding provided by the ozone layer.

8.   ----- The binomial system of naming organisms was developed by someone other than Whittaker.

9.   ----- The movement of Earth's crustal plates is thought to give rise to mountains.

10. ----- The mantle possesses a partially solid consistency.

11. ----- Molecular taxonomy utilizes analysis of proteins and DNA to construct taxonomic trees.

II.   Completion Exercise: Complete each sentence by filling in the blanks with the appropriate term or phrase. Write your answers in the space provided. Good luck!

1.   The formation of the Andes mountain range occurred due to _____.

2.   The three main geological eras are known as the _____, _____, and _____.

3.   Pangaea, the supercontinent, amalgamated from two separate landmasses: _____ and _____.

4.   According to Linnaeus' binomial system, each organism is identified by a dual name: the first indicating the _____, and the second denoting the _____.

5.   The broadest and most encompassing taxonomic category is the _____.

6.   The biological specification of a species relies on the concept of _____ isolation.

7.   Collections of organisms originating from a common ancestor are considered _____, or more precisely, stemming from a single clade.

III.   Multiple Choice: Choose the best answer for each question by circling the corresponding letter.

1. Most cosmologists believe that the universe originated _____.

a. 18 billion years ago

b. 4.5 billion years ago

c. 18 million years ago

d. 20 million years ago

e. 1.5 billion years ago

2. The prevailing theory regarding the inception of life on Earth suggests that _____.

a. Life arose spontaneously by chemical evolution

b. Life arose by physical evolution

c. Meteorites brought prokaryotic cells to earth

d. Life arose by natural selection

e. None of the above

3. Through replicating the conditions of early Earth, Urey and Miller demonstrated that _____.

a. Monomer formation can occur

b. Polymers can form

c. Organic molecules disintegrate

d. RNA replicates

e. All of the above

4. Scientists suggest that the initial polymers likely emerged _____.

a. In the earth's mantle

b. In primordial soup

c. In prehistoric oceans

d. On clay or rock surfaces

e. In microspheres

5. The potential source of energy driving organic polymerization might have been _____.

a.The sun

b.The earth's core

c.Ultraviolet light

d.ATP

e.All of the above

6. Based on the information presented in this chapter, life likely emerged on Earth _____.

a. About 5,000 years ago

b. Between 3 and 4 billion years ago

c. Between 2 and 3 billion years ago

d. About 500 billion years ago

e. About 10 billion years ago

7. Proteinoid microspheres can be generated _____.

a. By heating amino acids

b. By heating amino acids and exposing them to water

c. From a solution of polymers

d. From lipids

e. From coacervates

8. Ribozymes _____.

a. Are RN A molecules

b. Act as catalysts

c. Can build RNA molecules

d. Can carry out exchanges of RNA fragments

e. All of the above

9.   The most ancient fossils uncovered thus far are _____.

a. 2.5 billion years old

b. 5 billion years old

c. 3.5 billion years old

d. 1.5 million years old

e. 18 billion years old

10.   The initial cells were likely _____.

a. Aerobic

b. Anaerobic and autotrophic

c. Anaerobic and heterotrophic

d. Auxotrophic

e. Autotrophic

11.   As mentioned above, autotrophic cells _____.

a. Need oxygen to survive

b. Can survive without oxygen

c. Manufacture all their nutritional needs

d. Must ingest all required nutrients

e. a and d

12.   The development of an ozone layer in the Earth's atmosphere _____.

a. Enabled cells to survive in shallow water and on land

b. Resulted in the evolution of the first photo-synthetic plants

c. Resulted in the extinction of photosyn-thetic cells

d. Allowed the evolution of anaerobic cells to take place

e. Caused none of the above

13.   The cyclic movement of carbon compounds from the atmosphere to cells and back is termed _____.

a. Cellular respiration

b. The carbon cycle

c. Anaerobic respiration

d. Aerobic respiration

e. Continental drift

14.   It is hypothesized that eukaryotic organisms evolved _____.

a. 1.5 billion years ago

b. 4.5 billion years ago

c. 18 billion years ago

d. 1.5 million years ago

e. None of the above

15.   Plate tectonics, as a process, is primarily fueled by activity in _____.

a. The core

b. The crust

c. The mantle

d. The oceans

e. The atmosphere

16.   Liposomes play a crucial role in the formation of _____.

a. A lipid bilayer

b. Lipid storage

c. Microspheres

d. Monomers

e. None of the above

17.   Life originated during the _____.

a. Paleozoic

b. Mesozoic

c. Cenozoic

d. Cambrian

e. Precambrian

18. Reproductive isolation arises when _____.

a. Individuals of the same species live in different geographic areas

b. Individuals of separate species live on different continents

c. Two organisms cannot interbreed

d. Organisms are asexual

e. None of the above

19. Monophyletic organisms _____.

a. Share a common ancestor

b. Are reproductively isolated

c. Were the first forms of life

d. Live in shallow seas

e. None of the above

20. The study of how closely related groups diverged and branched apart is termed _____.

a. Generative biology

b. Classification

c. Taxonomies

d. Cladistics

e. Bioanalysis

READING UNDERSTANDING

Every autumn, countless green leaves burst into a dazzling array of reds, yellows, and browns, marking a striking seasonal transition. This change is the outcome of intricate chemical processes within the leaves. Essentially, the alteration in leaf color arises from the breakdown of chlorophyll, the pigment responsible for the leaves' green appearance. This breakdown occurs as the proteins holding the chlorophyll molecules disintegrate into amino acids. Consequently, the chlorophyll loses its green hue. The amino acids produced from this process are then transported through tiny sieve tubes to the stems and roots of the plant, where their nitrogen content is preserved for the upcoming season's growth.

Healthy chlorophyll functions by absorbing light across various colors, except for green, which it reflects, giving leaves their characteristic green appearance. However, as chlorophyll breaks down, the underlying yellow or brown hues of the leaf tissue become visible. Even during the summer months, chlorophyll undergoes partial breakdown while carrying out photosynthesis in sunlight, only to be replenished during the night. Consequently, leaves exhibit a subtle greening effect in the morning compared to the evening.

Despite botanists' comprehension of this process, the exact trigger or cause of autumn leaf color changes remains elusive. Speculations include the influence of cool weather, dehydration, long nights, and perhaps an internal biological timer, but definitive answers have yet to be uncovered.

1.   The primary focus of the passage is to explain___.

A. The mechanism of photosynthesis

B. The phenomenon of leaf color transformation

C. The function of chlorophyll in photosynthesis

D. The presence of atmospheric gaseous nitrates

2.   According to the passage, the proteins in leaves decompose and become ___ .

A. Amino acids

B. Chlorophyll

C. Sieve tubes

D. Botanists

3.   It can be inferred that proteins help maintain the green color of a leaf by ___ .

A. Producing new growth

B. Holding chlorophyll molecules

C. Breaking up into amino acids

D. Gathering nitrogen from the air

4. As stated in the passage, the color of the leaf tissue itself is __.

A. Green

B. Red

C. Yellow or brown

D. Primarily transparent

5. According to the passage, photosynthesis __ .

A. Relies on chlorophyll

B. Takes place exclusively during nighttime

C. Results in autumnal color changes of leaves

D. Could potentially be influenced by cool weather or extended nights

6. Which of the following queries cannot be addressed using the details provided in the passage?

A. What color of light does chlorophyll not absorb?

B. What color do leaves turn when lacking chlorophyll?

C. Which chemicals impart red and yellow hues to leaf tissue?

D. Is it possible for plants to retain nitrogen for later utilization?

READING MATERIALS

I. Earth: The Stage for Life

The exploration of space during the twentieth century stands as one of the most remarkable feats in human history. It offered profound insights into our solar system's central luminary, the sun, and the nine celestial bodies encircling it. Yet, as our understanding expands, a sense of solitude pervades, with many astronomers speculating that terrestrial life may be exclusive to Earth within our solar neighborhood. This prompts a fundamental question: Why did life emerge and thrive on our planet while seemingly faltering or failing to materialize on neighboring celestial bodies?

Expanding on the significance of space exploration, has not only deepened our comprehension of celestial phenomena but has also fueled existential inquiries regarding the uniqueness of Earth as a haven for life. As scientists delve deeper into the mysteries of astrobiology, probing the conditions conducive to life beyond our planet, they confront the enigma of Earth's exceptional suitability for hosting diverse forms of life.

In exploring this perplexity, considerations encompass a myriad of factors, including planetary composition, atmospheric conditions, and the presence of essential resources. Understanding the precise interplay of these elements holds the key to unraveling why Earth stands as a singular sanctuary for life amidst the desolate expanse of space.

1. How the Earth Formed: from Big Bang to Big Rock

According to prevailing astronomical theories, the inception of our universe traces back to the seminal event known as the Big Bang, estimated to have occurred approximately 18 billion years ago. This cataclysmic explosion gave rise to the entirety of matter in the cosmos, with hydrogen and helium atoms comprising a staggering 99 percent of its elemental composition. Remarkably, traces of this cosmic event persist to this day, as evidenced by the detection of residual cosmic radiation dispersed throughout the vast expanse of space, serving as enduring echoes of the Big Bang.

Within the confines of our solar system, clouds of hydrogen and helium encircling the sun gradually coalesced and cooled, heralding the formation of a multitude of celestial bodies, including planets, moons, and myriad smaller entities. Geologists, through meticulous study and analysis, have pinpointed the genesis of our own planet Earth to approximately 4.6 billion years ago, signifying a pivotal milestone in the cosmic chronicle of planetary evolution.

2. Origins of Earth: A Blak and Blue World

Following its formation, Earth is believed to have existed as an immense desolate sphere of ice and rock, lacking oceans and atmosphere. Nevertheless, within the planet's depths, the release of radioactive elements generated energy that, coupled with the force of gravity, gradually heated the rock, transforming it into a molten mass that remained in a state of warmth for millions of years. During this cooling process, a dense core composed of iron, nickel, and other heavy elements formed, surrounded by a liquid outer core, a substantial mantle of heated rock, and, on the surface, a solidified crust of dark igneous rock.

Frequently, molten rock erupted through this surface, constructing cinder cones that dotted the otherwise flat landscape, ejecting lava and emitting clouds of gases from the planet's interior. It was from these vapor clouds that the Earth's primordial atmosphere began to take shape; gravity retained a layer of gases comprising carbon dioxide, water vapor, nitrogen, hydrogen sulfide, and traces of methane, constituting the planet's inaugural air, devoid of free oxygen. Moisture within this atmosphere precipitated onto the steaming surface as rain, subsequently evaporating and precipitating once more, gradually cooling the surface and giving rise to a shallow global ocean. This ocean encompassed the entirety of the planet, interspersed only by the emergence of steeply rising black cones, continually spewing forth lava and gases.

Due to the presence of volcanic ash in the atmosphere, the sky likely exhibited a pale blue hue, akin to occasional modern-day occurrences; it would have also been adorned with clouds and frequently marked by the passage of flaming meteors and asteroids—small remnants of primordial rock hurtling through space. This juxtaposition of black and blue, characterized by an oxygen-free atmosphere, served as the stage upon which life first emerged.

3. *Earth's Favorable Position in the Solar System*

Life, as we currently understand it, relies on the presence of liquid water and carbon compounds, elements crucial for its emergence. Proximity to the sun has a profound impact on the conditions of celestial bodies, as evidenced by the scorching temperatures experienced by Mercury and Venus, the sun's immediate neighbors. On these planets, extreme heat renders water vapor and carbon compounds exist solely as inorganic gases, rendering them inhospitable to life as we know it.

Situated between Venus and Mars, Earth occupies a unique position in our solar system. While Mars may have once harbored flowing water and potentially supported life, it now stands barren, with its water frozen and carbon largely sequestered in rocks. In contrast, Earth boasts a remarkable combination of attributes, including its composition, geological dynamics, size, and distance from the sun. The presence of a molten core and volcanic activity has facilitated the release of carbon dioxide and water vapor into the atmosphere, where gravity ensnares these gases like a protective shroud. This atmospheric blanket, in turn, traps sufficient solar heat to maintain moderate surface temperatures and keep water in its liquid form—a key prerequisite for life to flourish on Earth.

4.   *The Ancient Earth and the Origins of Life's Ingredients*

A significant conundrum in the investigation of life's origins revolves around the accessibility of organic components essential for the earliest cellular structures. The fundamental inquiry persists: did the primordial Earth possess the requisite organic building blocks, and if so, what were their origins?

Recent astronomical revelations have unveiled vast reservoirs of organic molecules within the outer realms of our galaxy, suggesting a potential role in planetary formation. Furthermore, astronomers speculate that Earth may have harbored organic compounds from its inception, further enriching the primordial environment.

Another plausible source of organic materials could have been meteorite clusters, comprised of nondescript dark-gray stones, originating from the nascent stages of planetary formation. These meteorites, traversing space, occasionally impact Earth, with chemical analyses revealing the presence of nucleotide bases crucial for DNA and RNA formation. Some scholars postulate that significant meteorite bombardment during Earth's infancy may have contributed to the influx of organic compounds.

Moreover, it is conceivable that the raw materials necessary for life originated from molecules present in the early Earth's atmosphere, catalyzed by energy sources and environmental conditions prevailing at that time. This primordial stage set the scene for the unfolding drama of life to commence.

II. The Invisible Saga: From Molecules to Cells

The journey towards understanding the chemical genesis of life may have left its imprint within sedimentary layers solidifying into rocks around 4 billion years ago. Yet, the likelihood of unearthing this fossilized chronicle remains remote, given Earth's tumultuous early history characterized by relentless bombardment from colossal meteorites for the initial 800 million years. These catastrophic impacts likely resulted in the complete melting of the Earth's nascent crust, limiting the age of the oldest discoverable rocks to approximately 3.8 billion years.

However, what captivates and perplexes researchers is the discovery of cellular relics resembling early microbial life forms, akin to methane-producing bacteria, within rocks dating back nearly 3.5 billion years. Additionally, traces of cellular activity have been discerned in rocks as ancient as 3.8 billion years. Evidently, life manifested prior to this era, yet the tangible evidence of its inception has been lost to the ravages of time. Biologists are thus compelled to engage in conjecture, experimentation, and simulation to reconstruct the potential steps leading to life's emergence, acknowledging the scarcity of definitive proof. Herein lies one conceivable narrative outlining five pivotal stages that may have culminated in life's inception.

1. Polymerization: The Formation of Long Molecular Chains

Biologists posit that in ancient tide pools, seawater potentially harbored organic precursors such as amino acids or nucleotides. Through the process of solar evaporation, these compounds may have been concentrated, subsequently undergoing drying under the intense heat of the sun or freezing during nighttime, ultimately leading to the formation of polypeptides and RNA molecules. Experimental studies corroborate that under such extreme environmental conditions, subunits have the propensity to assemble into elongated chains, with clay or other minerals occasionally expediting these reactions.

2. Self-Replication: The Emergence of the RNA World

The prevailing consensus among biologists suggests that the initial nucleotide chains likely comprised straightforward, single-stranded RNA molecules. Experimental evidence indicates that RNA strands possess the capacity to self-replicate, with strands of RNA being capable of generating copies of themselves. When RNA is introduced into a solution containing the bases A, U, G, and C, complementary RNAs occasionally emerge. Subsequently, complementary copies of these new RNAs are produced, resulting in a second generation of copies that are identical to the original RNAs.

Furthermore, certain RNA molecules, known as ribozymes, exhibit the ability to excise segments of themselves, which then function as rudimentary enzymes, facilitating the cleavage and recombination of other RNA molecules. It is plausible that within tranquil tide pools or waterlogged clays, simple RNA genes formed and replicated themselves through this mechanism. Biologists often characterize this phase as "the RNA world," signifying the pivotal role of RNA in the primordial stages of life's emergence.

3.   The Evolution of Molecular Interactions: From RNA to Genetic Information Transmission

The transmission of genetic information from genes to proteins is a cornerstone of biological understanding. However, the transition from random RNA structures to information-carrying RNA structures, and their subsequent interaction with proteins, remains a profound mystery. This pivotal step can be likened to the coevolution of the chicken and the egg.

Biologists grapple with the enigma of how this crucial transition occurred. Observations reveal that amino acids and nucleotides in aqueous solutions have an inherent tendency to associate with one another. In pursuit of answers, there are even endeavors underway to explore deep-sea vents for evidence of such interactions occurring in contemporary settings.

Through millions of years of these interactions, specific nucleotide sequences likely acquired significance, representing distinct amino acid sequences. Both types of molecules gradually assumed evolutionary relevance, with their shapes and interconnected functions being shaped by natural chemical processes. As evidenced by modern organisms, this intricate dance culminated in the emergence of a stable gene storage form—DNA—alongside RNA and proteins, facilitating the storage and utilization of genetic information.

4.   Emergence of Cell-like Structures: The Protocell Hypothesis

Compelling evidence indicates that under specific conditions of temperature, dryness, and pH, polypeptides and phospholipids have the propensity to spontaneously assemble into minute spherical structures. Some biologists speculate that these spheres may have intermittently enclosed self-replicating assemblies of genes and proteins. Termed "protocells," these compartmentalized systems potentially underwent selection based on their overall performance, including factors such as accurate replication and efficient absorption of raw materials.

Furthermore, the universality of certain RNA molecule sequences across all contemporary organisms suggests a shared ancestry. Consequently, some biologists propose that all life forms trace their origins back to a common ancestor known as the progenote—a type of early cell believed to be the precursor to all life on Earth.

5. *The Emergence of Coordinated Cellular Activities*

Protocells gradually developed metabolic pathways, potentially spurred by competition among compartments for limited environmental resources. This competition likely led to the selection of enzyme series capable of transforming abundant materials into less available ones, thus evolving metabolic pathways capable of harnessing energy from chemical bonds and synthesizing new materials. As these activities became increasingly integrated and interdependent, protocellular compartments transitioned into indistinguishable biological entities—living cells—exhibiting fundamental characteristics such as order, adaptation, and reproduction.

The exact timing and location of this transition remain elusive. However, existing evidence strongly suggests that, given the conditions prevailing on early Earth over several hundred million years, and considering the basic physical and chemical properties of matter, the emergence of life units was not only plausible but inevitable. Following the cessation of Earth's crustal melting from meteoric bombardment, living cells evolved and proliferated through a natural and orderly sequence of events: from the origin of the universe to the formation of stars, planets, and the accumulation of organic molecules, polymer formation, polymer interactions, compartmentalization, and finally, the emergence of life.

CHAPTER 6

Fungi: The Mighty Decomposers

Abstract: Fungi, a diverse group of approximately 175,000 species, play critical roles in ecosystems as decomposers, parasites, and symbiotic partners. They exhibit a fundamental body structure composed of hyphae, which form a mycelium network facilitating nutrient absorption and growth. Hyphae may be septate or coenocytic, and fungi reproduce *via* spores, categorized as dispersal or survival spores. Fungi classification involves distinguishing between lower fungi (Chytridiomycetes, Oomycetes, Zygomycetes) and higher fungi (Ascomycetes, Basidiomycetes, Deuteromycetes), with each class exhibiting unique reproductive and morphological traits. Lichens represent a notable symbiotic relationship between fungi and algae, demonstrating resilience and nutrient acquisition abilities. Evolutionary insights suggest that fungi, as eukaryotes, share a common ancestor with animals and plants, highlighting their shared evolutionary history. While fungi and prokaryotes share some basic cellular features, fungi evolved within the eukaryotic lineage, making them more closely related to animals and plants than to prokaryotes.

Keywords: Fungal classification, Hyphae, Lichens, Mycelium, Spore reproduction.

UNDERSTANDING THE CHAPTER'S FOCUS: A ROADMAP

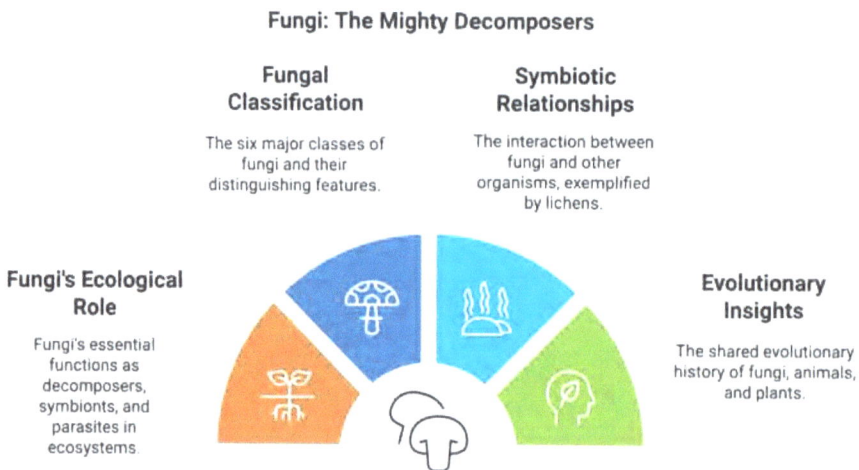

Fungi: The Mighty Decomposers

Fungal Classification
The six major classes of fungi and their distinguishing features.

Symbiotic Relationships
The interaction between fungi and other organisms, exemplified by lichens.

Fungi's Ecological Role
Fungi's essential functions as decomposers, symbionts, and parasites in ecosystems.

Evolutionary Insights
The shared evolutionary history of fungi, animals, and plants.

Mohammad Mehdi Ommati

1. FUNGI'S ATTRIBUTES

The vast array of fungi, comprising around 175,000 species, includes some of nature's simplest multicellular organisms. Exhibiting diverse lifestyles, fungi can function as **saprobes**, breaking down dead organic matter; they can adopt parasitic tendencies, drawing nutrients from living hosts; or they may form symbiotic relationships with algae or higher plant roots. Despite these varied roles, all fungi share a common trait: extracellular digestion. They secrete enzymes to break down organic material and then absorb the resulting nutrients.

Most fungi have a fundamental body structure consisting of a primary thallus made up of threadlike filaments called **hyphae** (Fig. **1**). The cell walls of hyphae typically contain chitin. In some species, hyphae specialize in forming rhizoids, which serve as root-like anchors, or transform into feeding structures known as **haustoria**. Additionally, hyphae can be either **septate** or non-septate, with septate hyphae having cross walls that divide individual cells, each containing at least one nucleus. Lower fungi exhibit a coenocytic structure, where the hyphae are a continuous mass of cytoplasm containing multiple nuclei.

Hyphae proliferate and branch out, forming a filamentous network called a **mycelium** (Fig. **1**). Digestion and nutrient absorption occur at the tips of each hypha, while new hyphae continue to form, enabling rapid fungal growth. This growth relies on mitosis and the rapid production of cytoplasm, with fungal mitosis occurring exclusively within the nucleus. Hyphae from genetically distinct organisms may fuse, resulting in a **heterokaryon**—one cytoplasm housing different nuclei.

As immobile heterotrophs, fungi must locate new nutrient sources over time. This critical function is carried out by spores, the reproductive entities of fungi. Spores are often borne on aerial hyphae and dispersed into the air. They can be either haploid or diploid, depending on the species. Spores fall into two primary categories: dispersal spores, which are short-lived and produced abundantly during active fungal growth, and survival spores, produced in smaller quantities during periods of environmental stress in the fungus's life cycle.

2. FUNGI CLASSIFICATION

Classifying fungi based on their evolutionary relationships can be challenging, as it requires an understanding of both molecular and morphological traits. Traditionally, fungi have been categorized by their morphology, reproductive methods, and spore production modes. Within the kingdom of *Fungi* (or *Mycota*), six primary classes are recognized.

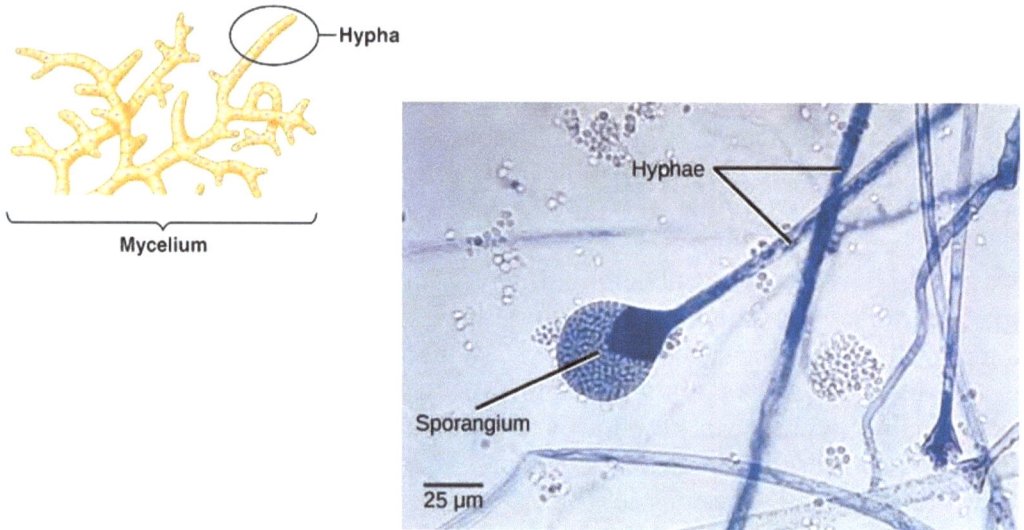

Fig. (1). Fungal Mycelia, Hyphae, and Sporangiospores. The right panel (the bright light micrograph) illustrates the release of spores from sporangium at the ends of hyphae called sporangiophores.

Lower fungi include *Chytridiomycetes*, *Oomycetes*, and *Zygomycetes*. These groups lack septate hyphae and are often coenocytic, relying primarily on asexual spore formation. Among these, only *Oomycetes* typically exhibit a diploid vegetative state. Both *Oomycetes* and *Chytridiomycetes* (commonly referred to as water molds) produce motile, flagellated spores in sporangia and gametes in **gametangia**. *Oomycetes* are distinctive for their large, immobile egg cells. Due to their unique characteristics, some biologists consider classifying these groups as protists rather than fungi. *Zygomycetes* resemble the previous two but produce nonmotile spores. Many *Zygomycetes* are terrestrial and some establish mycorrhizal relationships with plant roots.

Higher fungi include *Ascomycetes*, the largest class, where most members function as saprobes or parasites. Asexual reproduction produces **conidia**, which form at the tips of specialized aerial hyphae. During the sexual cycle, hyphae from different mating strains fuse, resulting in ascospores that are formed inside small, sac-like structures called asci. Fruiting bodies emerge from clusters of asci. Well-known ascomycetes include truffles, yeasts, and species of *Penicillium*.

Basidiomycetes, the second major group of higher fungi, typically form visible fruiting bodies. A key distinguishing feature is the dense mass of dikaryotic hyphae that make up the **basidiocarp**, often seen as mushrooms in moist environments. The **basidia**, club-shaped structures, bear four haploid basidiospores, which are released from the gills on the underside of the mushroom cap. *Basidiomycetes* undergo both sexual and asexual reproductive processes, depending on environmental conditions.

The class ***Deuteromycetes*** (or ***Fungi Imperfecti***) includes fungi that lack sexual reproduction mechanisms. Most reproduce asexually through the production of conidia. Human-relevant deuteromycetes include fungi used in the fermentation of soybeans and rice for soy sauce and sake, as well as those responsible for citric acid production and the toxic production of aflatoxins.

3. LICHENS: NATURE'S PERFECT SYMBIOSIS

Lichens are an epitome of symbiotic cooperation, where approximately 90 percent of the lichen structure consists of a fungal species, and the remaining 10 percent may harbor one or two types of algae. Within this partnership, the algae provide essential nutrients to the fungal counterpart, while the fungus potentially facilitates the algae's access to ecological niches otherwise inaccessible. While most lichen fungi belong to the *Ascomycetes* group, occasional occurrences of other higher fungi in lichens have been recorded. Lichens possess remarkable traits, such as the ability to survive near-complete desiccation without dying, and the capacity to absorb inorganic nutrients. Despite these extraordinary characteristics, the reproductive mechanisms of lichens remain an area of ongoing research.

4. FUNGAL EVOLUTIONARY INSIGHTS

Fungi, as eukaryotes, share a common ancestor with animals and plants, all of which belong to the eukaryotic domain. While fungi and prokaryotes share some basic cellular characteristics, fungi are more closely related to animals and plants than to prokaryotes. Molecular and genetic evidence supports the view that fungi diverged from a common ancestor with animals and plants long ago, positioning them more closely to these organisms on the evolutionary tree than to prokaryotes.

Fungi did not evolve directly from prokaryotic ancestors; instead, they are part of the eukaryotic lineage, which includes plants and animals. This evolutionary distinction highlights their more recent common ancestry with other eukaryotes, underscoring the greater evolutionary distance between fungi and prokaryotes.

Despite some functional similarities, such as basic metabolic processes, the shared eukaryotic traits in fungi, animals, and plants reflect a distinct evolutionary lineage separate from prokaryotes.

Vocabulary (Index of Terms)

Ascomycetes [ˌæskoʊˈmaɪsiːts]

Ascomycetes refer to a group of fungi characterized by the presence of a female reproductive structure called the ascocarp, where plasmogamy, the fusion of cytoplasm, occurs before the formation of the ascus.

Ascus [ˈæskəs] / asci [ˈæsaɪ]

The ascus is a distinctive structure found in Ascomycetes, responsible for producing sexual spores. These spores are essential for the reproduction and genetic diversity of the Ascomycete fungi.

Basidiocarp [baˌsidiauˈkaːp]

The basidiocarp is the prominent fruiting body of higher basidiomycete fungi. These structures, including mushrooms and toadstools, serve as the most conspicuous manifestations of these fungi's reproductive phase.

Basidiomycetes [baˌsidieuˈmaisiːtiːz]

Basidiomycetes are a group of mycelial fungi belonging to the Eumycophyta division. Their sexual reproduction involves the formation of basidia, typically borne on basidiocarps, except in the case of rusts and smuts.

Basidium [baˈsidium]

The basidium is the distinctive structure responsible for producing sexual spores in Basidiomycetes. These spores play a vital role in the reproductive cycle and genetic diversity of Basidiomycete fungi.

Conidium [keˈni,diam] (.i. conidia (kaˈni,die])

A conidium is an asexual spore produced by certain fungi, which is formed by abstriction from the tip or side of specialized hyphae called conidiophores. Conidia can be single-celled or multicellular, with septa arranged longitudinally, transversely, or in a mixed pattern in the case of multicellular conidia.

Fungi Imperfecti (Deuteromycetes) [ˈfʌˈ]gai imˈpeːfikti ([ˌdjuːteraumaiˈsiːtis])]

Fungi Imperfecti, also known as Deuteromycetes, represent a group of fungi characterized by septate mycelia, with either absent or unknown sexual reproduction. Many of these fungi hold significant importance as they are responsible for causing diseases in plants, animals, and humans.

Gametangium [ˌgremiˈtrendʒiəm]

A gametangium is a structure or organ in plants responsible for producing gametes, such as an antheridium, oogonium, or archegonium.

Haustorium [hoʊˈstɔːriəm]:

A haustorium is a specialized structure found in certain parasitic plants, such as dodder and mistletoe, which penetrates into the vascular bundles of the host plant to obtain nourishment.

Heterokaryon [ˌhetəroʊˈkæriən]

In somatic cell genetics, a heterokaryon refers to the state when two cells have fused, but their nuclei have not yet fused. Or it refers to a fungal cell containing two or more genetically distinct nuclei.

Hyphae [ˈhaɪfiː] (singular: hypha [ˈhaɪfə])

Hyphae are fungal filaments that emerge from the growth of germinating spores. They constitute the basic building blocks of a mycelium, the vegetative part of a fungus.

Lichen [ˈlaɪkən]

Lichen refers to thallophyte plants formed through a symbiotic relationship between a fungus and an alga. This symbiosis is so intimate that they create a vegetative plant body, morphologically distinct from either of the individual constituents.

Mycelium [ˈmaɪˈsiːliəm] - mycelia [maɪˈsiːliə]

Mycelium refers to the intricate network of hyphae that constitutes the vegetative body, also known as a thallus, of a fungus.

Rhizoid [ˈraɪzɔɪd]

A rhizoid is one of the various filamentous outgrowths, either uni- or multicellular, produced by certain algae and the gametophytes of bryophytes and pteridophytes. Rhizoids function similarly to roots, aiding in anchorage and absorption of nutrients.

Saprobe [ˈsrepraub]

An organism that lacks the ability of photosynthesis, and lives on dead or decaying organic matter, a saprophyte.

Septate [ˈsæproʊb]

When a fungal hypha is septate, it means it is divided by internal partitions called septa.

Sporangium [spəˈrændʒiəm]

A sporangium is a cell in which spores or spore-like bodies are produced. It is typically an asexual structure found in certain fungi.

COMPREHENSION TASKS

I. Key Terms: Matching

Instructions: Match each term from the left column with the corresponding description from the right column. Write the letter of the correct description next to the number of the term. Each description will only match with one term. Good luck!

1.	Rhizoid	**A.**	Having cross walls
2.	Ergot	**B.**	Cellular fungal filament
3.	Oomycetes	**C.**	A fungal feeding structure
4.	Mycorrhiza	**D.**	"Water molds"
5.	Aerial hypha	**E.**	A hyphal cell with two nuclei
6.	Conidium	**F.**	"Fungus root"
7.	Sporangium	**G.**	Spore sac
8.	Basidiomycetes	**H.**	Analogous to plant root
9.	Lichens	**I.**	Spore case
10.	Haustorium	**J.**	Network of hyphae
11.	Dikaryon	**K.**	Dustlike
12.	Mycelium	**L.**	Contains edible mushrooms
13.	Septate	**M.**	Composite organisms
14.	Ascus	**N.**	Source of drug LSD
15.	Hypha	**O.**	Tall spore-producing structure

Cell Biology True or False Quiz: Determine whether the following statements are true or false based on your understanding of cell biology concepts. Indicate 'True' for correct statements and 'False' for incorrect ones. Wishing you success!

1. ----- Fungal mitosis occurs within the nucleus, unlike in other organisms.

2. ----- The production of survival spores necessitates large numbers.

3. ----- Oomycetes are typically diploid.

4. ----- Zygomycetes are typically diploid.

5. ----- Dikaryons encompass some of the most detrimental fungal pests affecting humans.

6. ----- Oomycetes contain cellulose.

7. ----- Mycorrhizal associations are prevalent among 80 percent of terrestrial plants.

8. ----- The majority of basidiomycetes undergo asexual reproduction.

9. ----- Crustose lichens resemble miniature leaves.

10. ----- Despite their overall success, lichens struggle in arid climates.

11. ----- The fungal-to-algal ratio in lichens is approximately 9:1.

II.   **Completion Exercise: Complete each sentence by filling in the blanks with the appropriate term or phrase**. Write your answers in the space provided. Good luck!

1. Fungal chromosomes stand out from those of other eukaryotes due to their notably low levels of _____, which are essential proteins responsible for DNA coiling.

2. Plasmodara viticola, responsible for the devastating downy mildew outbreak in the French wine industry in 1880, belongs to the class _____.

3. The term for associations between fungi and higher plant roots is _____.

4. Approximately _____ percent of all terrestrial plants form these fungal associations.

5. Yeasts typically reproduce through _____.

6. The typical mushroom shape seen in "toadstools" and mushrooms of the Basidiomycetes class is actually a reproductive structure known as a _____.

7. Lichens possess an impressive capacity to remove airborne substances like sulfur and phosphorus, leading to their occasional use as _____ indicators.

8. Fungal cell walls incorporate _____ and, in certain species, _____.

III.   Multiple Choice: Choose the best answer for each question by circling the corresponding letter.

1. Fungi break down their food through the process of _____.

a.   Through digestion within their cells

b.   *Via* digestion between their cells

c.   Through digestion outside their cells

d.   With assistance from symbiotic organisms

e.   Not at all, as fungi are not capable of photosynthesis

2. Organisms that derive nutrients by decomposing dead organic matter are known as _____.

a.   Autotrophs

b.   Saprobes

c.   Protists

d.   Symbionts

e.   Parasites

3. The main body of a fungus is referred to as the _____.

a.   Haustorium

b.   Hypha

c.   Rhizoid

d.   Thallus

e.   Mycelium

4. Fungal haustoria can be described as _____.

a.   Feeding structures

b.   Reproductive organs

c.   Used for support

d.   The edible parts of mushrooms

e.   Produced in cycles

5. The composition of fungal cell walls primarily consists of _____.

a.   Cellulose

b.   Glycoprotein

c.   Glycolipid

d.   Chitin

e.   Inorganic compounds

6. Oomycetes and other lower fungi are classified as coenocytic due to their characteristic of _____.

a.   They lack hyphae

b.   They have a multinucleate cytoplasm

c.   They have septate hyphae

d.   They produce basidiospores

e.   Most are saprophytes

7. Fungi offer numerous benefits to human societies, but they do not typically provide _____.

a.   Food

b.   Medicines

c.   Fermenting agents

d.   Industrial filaments

e.   Commercial enzymes

8. Within fungi, the heterokaryon refers to _____.

a.   The fusion of two genetically distinct individuals

b.   The unique form of fungal mitosis

c.   The filamentous network of hyphae

d.   The one that arises by meiosis

e. None of the above

9. Effective genetic variation in fungi primarily emerges through_____.

a. Aerial hyphae

b. Crossing over

c. Heterokaryosis

d. Recombination

e. Rapid manufacture of cytoplasm

10. Dispersion spores typically possess the characteristic of _____.

a. Slow to germinate

b. Short-lived

c. Produced in low numbers

d. Made only in the spring

e. Endowed with thick cell walls

11. What role do commercial mushrooms play in the reproductive cycle of fungi?

a. By producing millions of spores

b. By forming symbiotic relationships with algae

c. By releasing enzymes for extracellular digestion

d. By undergoing sexual reproduction

e. By absorbing inorganic nutrients

12. Resting spores _____.

a. Are able to remain viable for long periods

b. Are formed when fungi experience slow growth rates

c. Are known to require a dormant phase be-fore germination

d. Are derived sexually

e. Are all of the above

13. Fungi have proven challenging to classify due to the fact that _____.

a. They are so similar in structure

b. They are such a diverse group

c. Most use the same means of nutrition

d. Their evolutionary relationships are un-known

e. Fungi are one of the easiest groups to classify

14. The fungal pathogen responsible for Dutch elm disease belongs to the class _____.

a. *Plasmopara viticola*

b. *Phytophora infestans*

c. *Rhizopus*

d. *Ophiostoma ulmi*

e. *Pseudomonas syringae*

15. Fill in the blank: Zygomycetes _____.

a. Can be found in both freshwater and saltwater environments

b. Are terrestrial

c. Are entirely autotrophic

d. Are never saprophytic

e. All of the above

16. What term describes the associations formed between plant roots and fungi?

a. Hyphae

b. Endosymbionts

c. Haustoria

d. Mycorrhizae

e. Mycelia

17. The class to which common mushrooms belong is _____.

a. Ascomycetes

b. Fungi Imperfecti

c. Zygomycetes

d. Basidiomycetes

e. Oomycetes

18. By what method do Deuteromycetes primarily reproduce asexually?

a. Basidia

b. A basidiocarp

c. Conidia

d. Ascospores

e. Asci

19. Roughly what proportion of a lichen's biomass consists of fungal tissue?

a. Ten %

b. Thirty %

c. Forty %

d. Eighty %

e. Ninety %

20. What sets lichens apart is their _____.

a. Their form of mitosis

b. Their ability to absorb inorganic nutrients

c. Their autotrophic lifestyle

d. Their lack of sexual reproduction

e. All of the above

READING UNDERSTANDING

I. Trees' Remarkable Survival Strategies

Trees stand out in the annals of Earth's history, boasting an unparalleled track record of resilience over an evolutionary span exceeding 400 million years. They have emerged as towering, robust, and enduring entities, unmatched by any other life form. Despite their formidable stature, trees lack a fundamental defense mechanism possessed by nearly all animals: mobility. Unlike their animal counterparts, trees are unable to escape from threats. As a consequence, they face a variety of threats, including natural elements like wildfires and storms, along with microorganisms, insects, various animal species, and ultimately humans, throughout their evolutionary development. Despite these challenges, trees have persevered due to their highly compartmentalized nature, effectively isolating and sealing off damaged or infected areas.

This distinct mode of defense sets trees apart from animals in a fundamental manner. Whereas animals possess the ability to heal, actively preserving their existence by undergoing countless reparations and replacing old cells with new or rejuvenated ones, trees follow a different strategy. Unlike animals, trees do not engage in repair mechanisms. Instead, they fend off the repercussions of injury and infection by encapsulating the affected areas. Concurrently, they initiate growth in new directions, essentially rejuvenating themselves annually by generating new cells and structures. The tangible evidence of this perpetual renewal process is evident in the formation of growth rings, prominently displayed on the cross-sections of trunks, roots, and branches.

In summary, trees' extraordinary survival mechanisms, characterized by their ability to compartmentalize damage and continually regenerate, underscore their resilience and longevity in the face of adversity.

1. What is the primary objective of the author in this article, focusing on elucidating the __?

A. The growth stages of a tree

B. How trees endure and thrive

C. The significance of trees in human advancement

D. Threats trees encounter from natural calamities

2. Except for which of the following traits does the author not attribute to trees?

A. High

B. Light green

C. Huge

D. Long-lived

3. The author suggests that nearly every animal can defend itself from harmful influences by __.

A. Departing

B. Requesting assistance

C. Ascending a tree

D. Staying with its group

II. Title: Unveiling the Intricacies of Food Webs and Their Components

In the intricate tapestry of ecosystems, the term "food web" emerges as a more apt descriptor of the complex interplay of feeding relationships than the linear notion of a "food chain." A food web represents a multifaceted feeding network composed of numerous interconnected food chains, each weaving together the dietary habits of various species. For instance, within a terrestrial ecosystem, the consumption dynamics unfold as mice, rabbits, and deer feed on plants, while owls prey on these herbivores. Anchoring these interactions is the foundational role of green plants, which, equipped with chlorophyll, harness energy from sunlight to produce sustenance.

In aquatic realms, algae serve as the primary producers, initiating food chains by synthesizing nutrients through photosynthesis. These microscopic green plants form the cornerstone of aquatic ecosystems, generating nourishment for a myriad of organisms. Tiny fish, sustained by algae, become prey for larger fish, thus perpetuating the flow of energy through successive trophic levels.

Organisms within ecosystems can be categorized into three primary groups based on their dietary habits: producers, consumers, and decomposers. Producers, exemplified by green plants and algae, utilize chlorophyll to manufacture organic compounds through photosynthesis. Consumers encompass a diverse array of species that derive sustenance from consuming other organisms, be they plants or animals. Decomposers, such as microbes, are essential for nutrient recycling as they decompose dead organic material. While producers and consumers contribute to the flow of energy within ecosystems, decomposers act as vital agents in the decomposition process, ultimately returning nutrients to the soil.

In essence, the intricate dynamics of food webs underscore the interconnectedness of life forms within ecosystems, highlighting the essential roles played by producers, consumers, and decomposers in sustaining ecological balance and biodiversity.

1. The primary aim of the passage is to ___ .

A. Identify the most effective food chain.

B. Depict the network of food interactions among flora and fauna.

C. Explain the process of photosynthesis in green plants.

D. Advocate for the safeguarding of endangered plant species to conservationists.

2. As per the author's perspective, how does the term "food web" signify the intricate interconnection of organisms within an ecosystem?

A. A complex arrangement comprising multiple interconnected food chains.

B. An organization responsible for food distribution.

C. The interconnectedness among various green plants.

D. The apparatus utilized by spiders to capture food.

3. What organism is typically positioned at the beginning of a food chain?

A. Wood-eating insects:Termites

B. Aquatic creatures

C. Lions

D. Plants like grass

4. The author categorizes organisms based on ___.

A. Their energy consumption

B. Their method of obtaining nourishment

C. The amount of energy needed for movement

D. Their habitat, whether terrestrial or aquatic

5. Which of the following organisms is least likely to act as a consumer, as described in the passage?

A. A microorganism

B. A rodent

C. A plant

D. An aquatic animal

READING MATERIALS

Exploring Microbiology: An Overview

Microbiology is frequently defined as the study of organisms and agents that are too small to be seen without the aid of a microscope, including various microorganisms such as viruses, bacteria, many algae, fungi, and protozoa. Despite this definition, some organisms within these categories, such as certain algae and fungi, are visible to the naked eye without needing a microscope. For instance, microbiologists examine visible organisms like bread molds and filamentous algae in their research. Interestingly, there are even bacteria, such as *Thiomargarita* and *Epulopiscium*, that are visible without magnification.

Establishing clear boundaries for microbiology has been difficult, which led Roger Stanier to suggest a more inclusive definition that considers both the size of the organisms studied and the methods utilized. Microbiologists usually start by isolating a particular microorganism from a larger population before cultivating it.

Consequently, microbiology employs a range of techniques, such as sterilization and the application of culture media, which are crucial for effectively isolating and nurturing microorganisms.

1. The Discovery of Microorganisms

Long before microorganisms became visible, early investigators speculated about their existence and their potential role in causing diseases. Notable figures like the ancient philosopher Lucretius (approximately 98-55 B.C.) and the doctor Girolamo Fracastoro (1478-1553) posited the idea that invisible living creatures might be responsible for illnesses. The first documented microscopic observations are credited to Francesco Stelluti, an Italian who studied bees and weevils from 1625 to 1630 using a microscope that was likely supplied by Galileo.

Nevertheless, it was Antony van Leeuwenhoek (1632-1723), an amateur microscopist from Delft, Netherlands, who was the first to accurately observe and characterize microorganisms. Despite being a draper and haberdasher by trade, Leeuwenhoek dedicated a significant portion of his free time to designing basic microscopes, which featured double convex glass lenses positioned between two silver plates. These primitive microscopes were capable of magnifying objects by about 50 to 300 times. Leeuwenhoek probably used a type of illumination technique by sandwiching his liquid samples between two glass slides and shining light on them at a 45° angle, which made the bacteria distinctly visible. From 1673 onward, Leeuwenhoek diligently communicated his findings through detailed letters to the Royal Society of London. His descriptions leave no doubt that he observed both bacteria and protozoa during his investigations.

2. The Controversy Surrounding Spontaneous Generation

Historically, the belief in spontaneous generation—the notion that living beings could arise from nonliving material—was widely accepted. Even prominent thinkers such as Aristotle (384-322 B.C.) theorized that some basic invertebrates could come into existence this way. This idea was later questioned by the Italian physician Francesco Redi (1626-1697), who performed a series of trials with decaying meat to examine the supposed spontaneous generation of maggots.

In Redi's experiments, he placed the meat in three different containers: one left open, another sealed with paper, and the last one shielded with thin gauze to prevent flies from accessing the meat. Maggots only developed on the uncovered meat where flies could access it directly. The sealed containers did not generate larvae on their own. However, flies were drawn to the gauze-shielded container, laying eggs on the gauze, which eventually led to the appearance of maggots. These experiments demonstrated that maggots did not spontaneously generate from decaying meat as previously believed but rather from fly eggs.

Although spontaneous generation had been dismissed for more complex organisms, Leeuwenhoek's findings of microorganisms brought the debate back into focus. Some scientists argued that while larger organisms did not arise spontaneously, microorganisms might. They referenced experiments in which boiled hay or meat extracts eventually led to the emergence of microorganisms. In 1748, English priest John Needham reported the results of his experiments, boiling mutton broth and observing microbial growth in tightly stoppered flasks. He suggested that organic substances possessed a vital force that could transfer life-giving properties to nonliving substances. Later, the Italian clergyman and naturalist Lazzaro Spallanzani enhanced Needham's trials *via* sealing flasks filled with water and seeds, demonstrating that no growth occurred when the sealed flasks were boiled.

The debate continued with various investigators attempting to counter arguments for spontaneous generation. Theodore Schwann permitted air to flow into a flask containing a sterile nutrient solution by directing it through a red-hot tube, and the flask stayed sterile. Similarly, Georg Friedrich Schroder and Theodor von Dusch permitted air to pass through heat-sterilized media by filtering it through germ-free cotton wool, causing the medium to remain free of growth. However, in 1859, French naturalist Felix Pouchet claimed to have proven microbial growth without air contamination. Louis Pasteur countered this claim, demonstrating that air filtered through cotton contained trapped objects resembling plant spores that could lead to microbial growth when introduced to a sterile medium. Pasteur's experiments with curved-neck flasks further solidified his argument against spontaneous generation, showing that dust and microorganisms caught on the curved necks prevented microbial growth in boiled solutions.

By 1861, Pasteur had effectively resolved the controversy and established methods for maintaining sterile solutions. English physicist John Tyndall reinforced this idea in 1877 by showing that dust carried germs and that a sterile broth stayed uncontaminated when exposed to air, provided no dust was present. Independently, German botanist Ferdinand Cohn identified bacterial endospores capable of surviving high temperatures, further supporting the case against spontaneous generation.

3. The Contribution of Microorganisms to Illness

Initially, the significance of microorganisms in causing illness was not readily apparent to the public, and a considerable amount of time passed before researchers established a connection between microorganisms and disease. This understanding hinged heavily on the advancement of new techniques for studying microorganisms. Once it was established that microbial infections could indeed lead to disease, microbiologists started investigating how hosts defended themselves against these microorganisms and explored methods for disease prevention. This exploration laid the foundation for the field of immunology.

3.1. Understanding the Connection Between Microorganisms and Disease

Although some early theorists, such as Fracastoro, proposed that invisible organisms could cause disease, the prevailing beliefs attributed illnesses to supernatural influences, miasmas (toxic gases), and disruptions in the body's four humors. This notion of humoral imbalance, going back to the time of Galen, was widely accepted. However, backing for the germ theory of disease started to increase in the early 1800s.

In 1835, Agostino Bassi reported that a fungal infection caused a disease in silkworms, marking the first instance of proving a microorganism's role in disease. He further proposed that many diseases could be attributed to microbial infections. Additionally, M.J. Berkeley established in 1845 that a fungus was responsible for the Great Potato Blight in Ireland.

Pasteur, famous for his research on fermentation, was commissioned by the French government to study the pebrine disease plaguing silkworms. After extensive research, he determined that a protozoan parasite was responsible for the disease and suggested that it could be managed by selecting healthy moths for breeding.

Joseph Lister, an English surgeon, provided indirect evidence supporting the involvement of microorganisms in human diseases through his work. Building on Pasteur's discoveries regarding fermentation and decay, Lister pioneered antiseptic surgical methods to inhibit the entry of microorganisms into wounds. This process involved sterilizing instruments with high heat and using phenol on wound dressings and as a mist around the operative site. Lister's method transformed the field of surgery following the publication of his observations in 1867. He offered convincing proof of the connection between microorganisms and illness, demonstrating that phenol eliminated bacteria and, in turn, prevented infections in wounds.

A key breakthrough in proving bacteria as the cause of disease came from German physician Robert Koch's research on anthrax. Building on principles outlined by his mentor, Jacob Henle, Koch demonstrated the connection between *Bacillus anthracis* and anthrax, publishing his results in 1876. He introduced infected material from sick animals into healthy mice, which subsequently fell ill. Through a series of 20 mouse inoculations, he successfully transferred anthrax. Subsequently, he cultured anthrax bacilli from spleen tissue in beef serum, observing their growth, reproduction, and spore production. Anthrax was induced in mice by introducing isolated bacilli or spores. These experimental steps laid the foundation for what became known as Koch's postulates, which are summarized as follows:

- The microorganism must be present in every case of the disease but absent from healthy organisms.

- The suspected microorganism must be isolated and grown in a pure culture.

- Inoculating the isolated microorganism into a healthy host must produce the same disease.

- The same microorganism must be isolated again from the diseased host.

Although Koch utilized comparable techniques in his studies on anthrax, he did not explicitly define these postulates until his publication on tuberculosis in 1884. His revolutionary research on Bacillus anthracis was later validated by Pasteur and his colleagues, who found that anthrax spores could survive when the remains of dead animals were buried in the ground and were subsequently unearthed by earthworms. When healthy animals consumed these spores, they became ill, thereby supporting Koch's conclusions.

3.2. Advances in Methods for Investigating Microbial Pathogens

Throughout Robert Koch's research on bacterial infections, the need arose to isolate potential disease-causing bacteria. Initially, Robert attempted to culture these prokaryotes on cut, boiled potatoes, but this method proved unreliable as bacteria didn't always grow well on potatoes. Koch then experimented with solidifying liquid media using gelatin. This allowed distinct bacterial colonies to form, either by spreading a bacterial sample across the surface or by mixing the sample into the liquefied gelatin medium, which would harden to form individual bacterial colonies. However, gelatin had its limitations— It was susceptible to bacterial degradation and would melt at temperatures exceeding 28°C.

A significant breakthrough came from Fannie Eilshemius Hesse, the spouse of Walther Hesse, who was one of Koch's assistants when she suggested using agar as a solidifying agent. Having successfully used agar to make jellies, she found it resistant to bacterial digestion and stable up to 100°C. Koch's assistant, Richard Petri, further advanced the process by creating the Petri dish, a vessel designed for solid culture media. These advancements facilitated the separation of uncontaminated bacterial cultures that contained a single species, significantly advancing the field of bacteriology.

Koch also formulated media designed for cultivating bacteria sourced from the body, using meat extracts and protein breakdown products due to their resemblance to bodily fluids. Koch had utilized these methods to successfully separate the bacillus responsible for tuberculosis in 1882, marking the beginning of a remarkable golden era that lasted approximately three to four decades. During this period, significant advancements were made in microbiology, leading to the discovery of the majority of major bacterial pathogens, which fundamentally transformed our understanding of infectious diseases and paved the way for modern bacteriology and public health interventions.

The identification of viruses and their involvement in disease was advanced by Charles Chamberland, a collaborator of Pasteur, who developed a porcelain bacterial filter in 1884. This development paved the way for studying virus-related pathogens, with the tobacco mosaic disease virus being the first to be investigated.

3.3. Advancements in Immunological Research

Significant strides were made during this time in understanding how animals combat diseases, alongside the development of methods to safeguard both humans and livestock from pathogens. Louis Pasteur and Emile Roux made key discoveries during their studies on chicken cholera. They discovered that prolonged incubation of bacterial cultures during transfers weakened the bacteria, making them incapable of causing disease. Injecting chickens with these attenuated cultures conferred immunity against the disease, a process Pasteur termed "vaccination" in homage to Edward Jenner, who had pioneered vaccination against smallpox using cowpox material.

Building on this achievement, Pasteur and Chamberland formulated an attenuated vaccine for anthrax, utilizing methods such as potassium bichromate treatment and specific temperature incubation of the bacteria. Pasteur adopted a different approach to developing[1] the rabies vaccine. The pathogen was weakened by cultivating it in an unusual host, the rabbit. Once the rabbits succumbed to the infection, their brains and spinal cords were carefully dissected and subsequently fixed. Notably, Pasteur successfully treated a nine-year-old boy named Joseph Meister, who had been bitten by a rabid dog, by administering progressively more virulent doses of the attenuated virus. Grateful for Pasteur's contributions to vaccine development, people worldwide supported the establishment of the Pasteur Institute in Paris, which initially specialized in the production of vaccines.

The identification of the toxin produced by the diphtheria bacillus resulted in further advancements by Emil von Behring and Shibasaburo Kitasato. They administered inactivated toxin to rabbits, prompting them to generate antitoxins—substances in the blood capable of neutralizing the toxin and providing protection against the disease. This research demonstrated that immunity could be generated by soluble components present in the blood, now known as antibodies (humoral immunity). Additionally, Elie Metchnikoff found that specific blood leukocytes were capable of engulfing disease-causing bacteria, a process Elie termed *phagocytosis*, emphasizing the importance of cellular immunity.

4. Microbial Applications in Industry and Ecosystem Dynamics

In 1837, Theodore Schwann and his colleagues suggested that yeast cells were responsible for transforming sugars into alcohol (ethanol) and carbon dioxide, a process referred to as alcoholic (ethanol) fermentation. However, prominent

---

[1] • **To develop**: This suggests that the action is about the entire process of creating the vaccine.

  • Example: "Pasteur took an alternative method to develop the rabies vaccine."

• **To developing**: This suggests a focus on the process itself, as if emphasizing the method of preparation.

  • Example: "Pasteur took an alternative approach to developing the rabies vaccine."

chemists of that era disagreed, attributing fermentation to chemical instability rather than microbial activity. Louis Pasteur challenged this notion, believing that fermentation was conducted by viable microorganisms and resulted in asymmetric compounds with optical activity, including amyl alcohol. In 1856, Pasteur was approached by M. Bigo, an industrialist in Lille, France, seeking assistance with ethanol production from beet sugars. Pasteur found that problems in fermentation occurred when yeast, which is crucial for alcohol production, was substituted by microorganisms that generate lactic acid. This practical problem-solving led Pasteur to demonstrate that fermentations were orchestrated by distinct yeasts and bacteria, resulting in multiple publications on the topic between 1857 and 1860. His research expanded to include investigations into wine spoilage and led to the invention of pasteurization as a method to preserve wine during storage. Over nearly 20 years, Pasteur explored various aspects of fermentation, including the discovery that certain microorganisms involved in the process were anaerobic, meaning they could only survive in oxygen-free environments.

Meanwhile, a few microbiologists delved into the ecological roles of microorganisms, particularly in the nitrogen, carbon, and sulfur cycles within soil and aquatic ecosystems. Sergei N. Winogradsky, a pioneering Russian microbiologist, made notable advancements in soil microbiology. He discovered that certain soil bacteria could derive energy by oxidizing iron, sulfur, and ammonia, and showed that many bacteria could assimilate CO_2 into organic compounds, much like photosynthetic organisms. Additionally, Winogradsky isolated anaerobic nitrogen-fixing bacteria from soil and explored the processes involved in cellulose degradation.

Martinus W. Beijerinck, another prominent microbiologist, made crucial advancements in microbial ecology. He was responsible for isolating bacteria that fix nitrogen aerobically, such as Azotobacter, as well as sulfate-reducing microorganisms. Along with Winogradsky, he introduced the enrichment-culture method and pioneered the application of selective media, which were groundbreaking developments in microbiological research.

5. *Exploring Microbial Diversity: Prokaryotes, Eukaryotes, and Modern Classification*

Microbiologists study a diverse array of organisms, which can be broadly categorized into two types of cells: prokaryotic and eukaryotic. Prokaryotic cells do not possess a well-defined, membrane-bound nucleus and exhibit simpler morphology, making up all bacteria. In contrast, eukaryotic cells contain a membrane-bound nucleus, have a more intricate structure, and are generally larger than prokaryotic cells. Eukaryotes encompass organisms such as algae, protozoa, fungi, vascular plants, and animals. Prokaryotic and eukaryotic cells vary in various aspects beyond morphology.

Historically, the classification of organisms encompassed five kingdoms: Monera, Protista, Fungi, Animalia, and Plantae. Microbiologists generally focus their research on organisms belonging to the first three kingdoms. Moreover, microbiologists also examine viruses, which, while excluded from the conventional five-kingdom classification, are significant subjects of study.

Recent advancements have significantly impacted microbial classification. Electron microscopy has provided insights into the detailed structure of microbial cells. Microbiologists have investigated the biochemical and physiological properties of a wide range of microorganisms. Furthermore, comparing nucleic acid and protein sequences from various organisms has shed light on microbial diversity.

It is recognized that prokaryotic organisms can be categorized into two separate groups: Bacteria and Archaea. Moreover, the extensive variety within protists has led some researchers to suggest dividing the kingdom Protista into several distinct kingdoms. Consequently, numerous taxonomists contend that the conventional five-kingdom system is overly simplistic and advocate for alternative classification schemes. The distinctions among Bacteria, Archaea, and eukaryotes are significant enough that some microbiologists propose classifying organisms into three domains: Bacteria (also known as true bacteria or eubacteria), Archaea, and Eucarya (encompassing all eukaryotic life forms).

6. Microbiology: Its Scope and Importance

As Steven Jay Gould insightfully pointed out, we are presently in the Age of Bacteria. These organisms, the earliest forms of life on Earth, inhabit virtually every possible environment, outnumbering all other life forms and likely constituting the largest portion of the Earth's biomass. Their activities are integral to the functioning of ecosystems and profoundly impact human society in plentiful routes. Contemporary microbiology encompasses a broad field with various specializations, influencing areas such as healthcare, agricultural practices, nutrition science, environmental science, hereditary studies, biochemical sciences, and cellular biology. Microbiology has been instrumental in the advancement of molecular biology, emphasizing the physical and chemical properties of living organisms and their functions. Microbiologists have made significant contributions to research on the genetic code, as well as the processes involved in DNA, RNA, and protein synthesis. Microorganisms were essential in the foundational research on gene expression regulation and enzyme activity control, which eventually facilitated advancements in recombinant DNA techniques and genetic manipulation in the 1970s.

The importance of microbiology during the 20ᵗʰ century is highlighted by the fact that roughly one-third of the Nobel Prizes in Physiology or Medicine were awarded for discoveries in microbiological research.

Some microbiologists focus on the detailed biology of microorganisms, specializing in areas such as virology, bacteriology, mycology, and microbial physiology, as well as ecology, genetics, and molecular biology. Others utilize their expertise to address practical challenges in areas such as medical microbiology, food and dairy microbiology, and public health microbiology, where research meets real-world applications. For instance, medical microbiologists identify infectious disease agents and develop strategies for their elimination, while public health microbiologists focus on preventing and controlling the transmission of infectious diseases and ensuring the safety of food and water supplies.

Immunology, a rapidly growing field, explores how the immune system defends against pathogens and addresses health issues such as allergies and autoimmune diseases. Agricultural microbiology aims to enhance crop yields, combat plant diseases, and explore alternatives to synthetic pesticides. Microbial ecology investigates the association between microorganisms and their habitats, contributing to understanding carbon, nitrogen, and sulfur cycles and developing strategies for pollution control through bioremediation.

Food and dairy microbiologists focus on preventing food spoilage and the spread of foodborne diseases, while harnessing microorganisms for the production of diverse food items.In industrial microbiology, microorganisms are employed to manufacture antibiotics, vaccines, steroids, alcohols, enzymes, and other valuable substances. Research in microbial physiology, biochemistry, genetics, and molecular biology delves into topics such as antibiotic synthesis, microbial energy production, nitrogen fixation, and genetic engineering, offering insights into gene function and the development of novel microbial strains for practical applications in medicine, agriculture, and industry.

7. The Future of Microbiology

Microbiology holds great promise for the future, driven by its clear mission and practical significance. As new challenges emerge, microbiologists are poised to address them through innovative research and technological advancements. Here are some emerging research fields and their possible practical applications, along with the challenges faced by microbiologists:

1. **Emerging Infectious Diseases:** Continuously arising new infectious diseases and the resurgence of old ones pose significant threats. Microbiologists must respond to these challenges by identifying and developing strategies to combat unknown pathogens.
2. **Antibiotic Resistance:** Addressing the increasing antibiotic resistance of pathogens is critical. Microbiologists must create alternative treatments and devise strategies to curb or halt the progression of drug resistance, employing methods from molecular biology and genetic engineering.
3. **Chronic Diseases and Infections:** Further research is required to explore the link between pathogenic organisms and long-term conditions such as autoimmune disorders and cardiovascular diseases. Some chronic conditions might be influenced by infections.
4. **Host-Pathogen Interactions:** Understanding how pathogens interact with host cells and how diseases arise is crucial. Microbiologists must explore host resistance mechanisms and disease development pathways.
5. **Industrial and Environmental Applications:** Microorganisms play vital roles in industry and environmental control. Research is required to harness their potential in food production, pollutant degradation, disease treatment, and agricultural productivity enhancement.
6. **Microbial Diversity:** Exploring microbial diversity, particularly in extreme environments, is essential. Developing new isolation techniques and classification methods will drive progress in manufacturing technologies and ecological management.
7. **Biofilms and Microbe-Microbe Interactions:** Biofilms, communities of microorganisms, have significant implications in medicine and ecology. Research on biofilms and microbe-microbe interactions is nascent but holds promise for practical applications.
8. **Genomics and Bioinformatics:** Advancements in genome sequencing and bioinformatics enable insights into microbial cell structure and function. Analyzing genomes will require continued progress in bioinformatics and computational biology.

9. **Symbiotic Relationships:** Understanding symbiotic relationships between microorganisms and higher organisms is crucial. This knowledge can improve ecosystem health, agricultural practices, and human health outcomes.

10. **Fundamental Biological Questions:** Microorganisms serve as excellent models for studying fundamental questions in biology, such as cellular development, communication, and environmental responses.

11. **Ethical Considerations:** Microbiologists must carefully consider the ethical implications of their research and technological developments, balancing societal benefits with potential risks.

Despite the challenges, the future of microbiology is bright, offering exciting opportunities to contribute to the betterment of human life. Microbiologists play a crucial role in various scientific disciplines and have the potential to address pressing global issues while advancing our understanding of the microbial world.

CHAPTER 7

The Journey of Animal Development: From Gamete Production to Fertilization

Abstract: Animal development begins with the production of gametes through spermatogenesis and oogenesis, processes crucial for sexual reproduction. Spermatogenesis produces sperm in the testes, characterized by its tail and acrosome, while oogenesis forms eggs in the ovaries, which vary in size and complexity. Fertilization, whether internal or external, initiates development by merging male and female gametes to form a diploid zygote. The subsequent cleavage stage involves rapid cell division, forming a blastula, and setting the stage for gastrulation, where the embryo develops distinct germ layers. Organogenesis follows, leading to the specialization of cells into functional tissues and organs. In land vertebrates, extraembryonic membranes protect and nourish the embryo, while growth dynamics are driven by cellular proliferation rather than individual cell enlargement. Aging concludes the developmental journey, characterized by a structural and functional decline over time.

Keywords: Cleavage, Fertilization, Gametogenesis, Gastrulation, Organogenesis.

UNDERSTANDING THE CHAPTER'S FOCUS: A ROADMAP

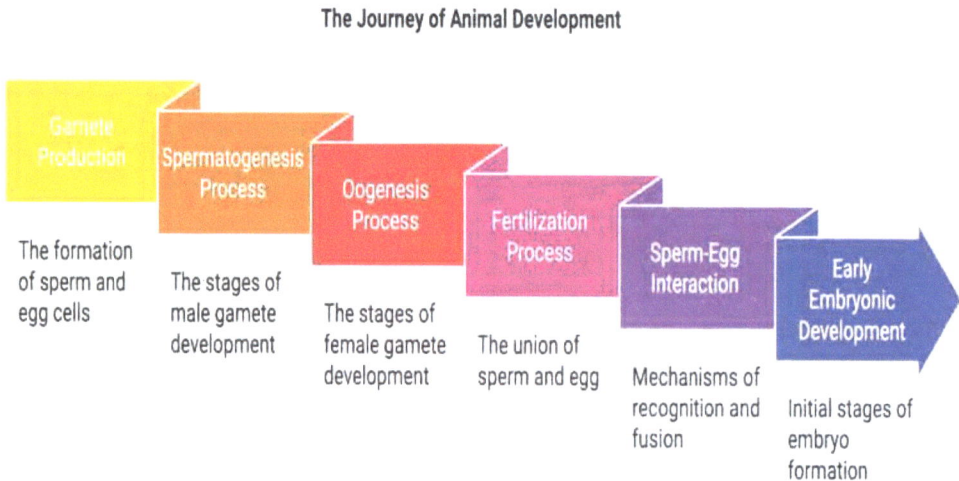

The Journey of Animal Development

Gamete Production — The formation of sperm and egg cells

Spermatogenesis Process — The stages of male gamete development

Oogenesis Process — The stages of female gamete development

Fertilization Process — The union of sperm and egg

Sperm-Egg Interaction — Mechanisms of recognition and fusion

Early Embryonic Development — Initial stages of embryo formation

1. GENERATING SPERM AND EGGS

In organisms that reproduce sexually, males and females produce specialized sex cells called gametes, which are **sperms** in males and eggs (**ova**) in females. These processes in males and females are called gametogenesis, including spermatogenesis in males and oogenesis in females (Fig. **1**). **Spermatogenesis**, the process of sperm production, occurs in the testes, originating from gonial cells within seminiferous tubules. Through mitosis and meiosis, spermatocytes develop into haploid spermatids, each equipped with a nucleus containing chromosomes, a tail, and an acrosome housing enzymes crucial for fertilization. Conversely, **oogenesis**, the formation of eggs, begins with oogonia in the **ovaries**, undergoing a series of meiotic divisions after maturation and ovulation, resulting in the development of a fertilizable egg.

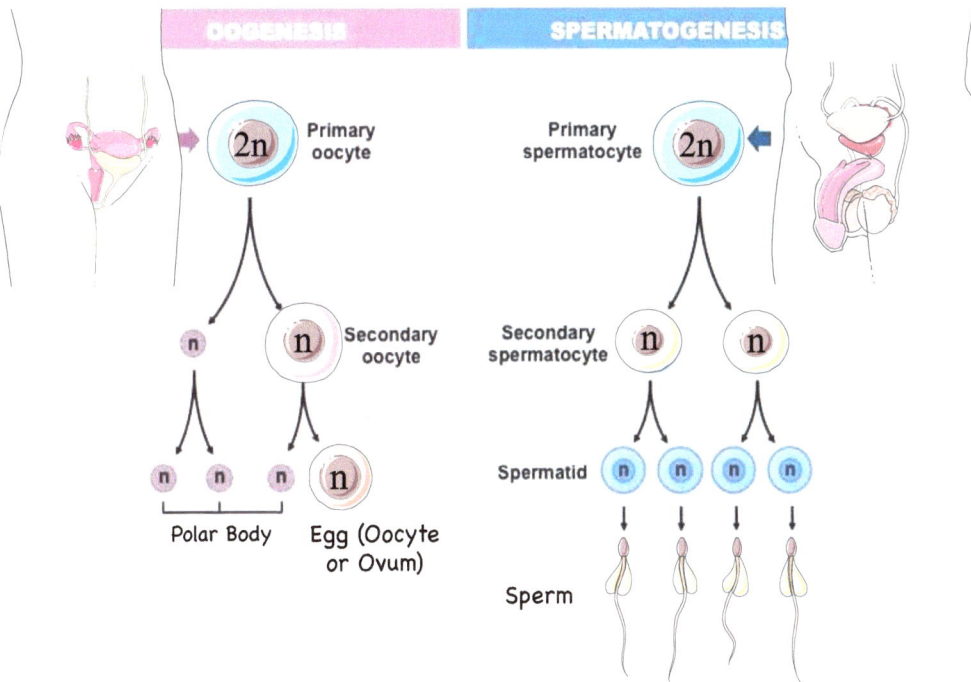

Fig. (1). Gametogenesis (Spermatogenesis and Oogenesis).

Eggs exhibit considerable variation in size and complexity across species and are typically enveloped by supporting cells such as follicle cells or nurse cells. They also store varying amounts of **yolk**, sourced from the mother's digestive glands, serving as a nutrient reservoir for the developing embryo. Additionally, eggs are shielded by protective coatings, including albumen and outer membranes, provided by follicle cells or maternal **oviduct** cells.

Frog oocytes have emerged as invaluable model systems for investigating oocyte development. Throughout maturation, these oocytes undergo a remarkable process of ribosome production facilitated by **gene amplification**, leading to the generation of vast quantities of ribosomes. Additionally, substantial amounts of mRNA are synthesized and stored during this period, further contributing to the intricate regulatory mechanisms governing oogenesis in frogs.

2. FERTILIZATION: THE GENESIS OF DEVELOPMENT

Fertilization, the union of male and female gametes, marks the inception of development. This process can occur either externally or internally, with the initial sperm-egg interaction triggering the **acrosome reaction**. This reaction involves the release of enzymes to facilitate sperm penetration through the egg's protective layers. Upon fusion of the sperm and egg plasma membranes, the male nucleus enters the egg cytoplasm, merging with the egg nucleus to form a diploid **zygote** (Fig. **2**). The egg's cortical reaction prevents polyspermy by ensuring that only one sperm fertilizes the egg, with the elevation of the fertilization membrane further impeding additional sperm entry.

In some species, such as reptiles, insects, and some invertebrates, fertilization can be bypassed through a process known as **parthenogenesis**. In this process, the egg activates and embarks on embryonic development without external fertilization, leading to offspring development. Parthenogenesis can produce both male and female offspring, depending on the species and the type of parthenogenesis exhibited. This reproductive strategy highlights the diversity and adaptability of reproduction across the animal kingdom.

Fertilization occurs on Day 0 when the sperm and egg unite to form a zygote. Following fertilization, the egg completes its second meiotic division, producing a polar body, and the zygote undergoes its first mitotic division, known as cleavage. These early divisions are rapid and lead to the formation of smaller cells, marking the initial stages of development (Fig. **2**).

3. CLEAVAGE: THE FOUNDATION OF CELL PROLIFERATION

Cleavage, the pivotal stage following fertilization, represents a specialized form of cell division (mitosis). It results in the formation of a **blastula**, characterized by a spherical arrangement of cells surrounding a central cavity, which is a hallmark of early embryonic development in many species. During cleavage, the zygote divides, generating numerous smaller cells, each with a complete diploid set of chromosomes (Fig. **2**). Essential materials such as yolk, mRNA, and ribosomes are distributed to ensure uniform cellular development.

The pattern of cleavage varies among species, primarily influenced by the amount of yolk present in the egg. In mammals, which have minimal yolk, cleavage occurs uniformly, resulting in cells of roughly equal size. Conversely, species like frogs, with moderate yolk content, show more rapid cleavage in areas with less yolk. Birds, with their extensive yolk reserves, undergo limited cleavage divisions confined to a small cytoplasmic region.

In many species, the accurate allocation of molecular determinants within **blastomeres** is crucial for directing subsequent development. In mammals and birds, the positional orientation of cells during cleavage plays a critical role in determining cell fate, underscoring the importance of spatial organization in forming diverse cell types within the embryo.

Summary of Cleavage in Early Development

On **Day 0**, the fertilized egg undergoes its first mitotic division, resulting in a zygote and the formation of the first polar body. By **Day 1**, the zygote reaches the 2-cell stage, and by **Day 2**, it progresses to the 4-cell stage. On **Day 3**, the embryo reaches the early morula stage, a solid ball of cells. By **Day 4**, the morula becomes more differentiated, progressing into the late morula stage. By **Day 6**, the blastocyst forms, including the blastocoele (blastocyst cavity), trophoblast, and inner cell mass (ICM). The blastocyst is crucial for implantation, which typically occurs between **Days 7-10** in species like humans and mice. The embryo embeds into the uterine wall, where further differentiation and development continue, though the timing can slightly vary between species (Fig. **2**).

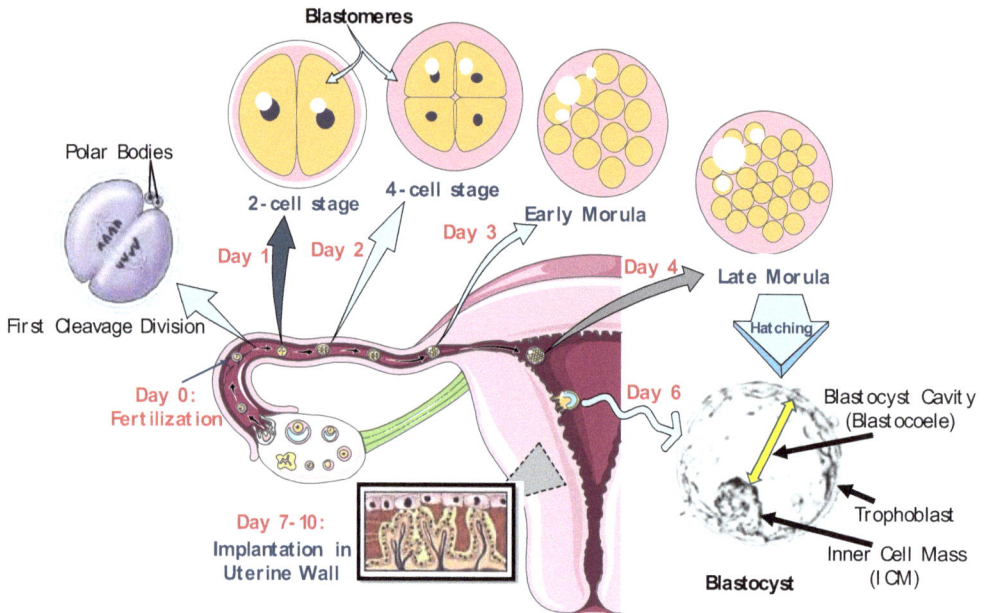

Fig. (2). Embryonic development stages from fertilization to implantation in humans and mice: Cleavage, morula formation, and blastocyst generation.

Species-Specific Differences in Embryonic Development

Embryonic development across different species follows broadly similar stages, from fertilization to implantation, yet considerable variation exists in the timing and specifics of these events (Table **1**). These differences are primarily influenced by factors such as the mode of fertilization, the yolk content in the egg, and the particular reproductive strategies of each species. Understanding these variations can provide valuable insights into comparative embryology, as well as evolutionary and developmental biology.

1. *Fertilization and Early Cleavage*

Fertilization marks the beginning of embryonic development. It typically occurs on Day 0, when the sperm and egg unite to form a zygote. This process initiates the first cleavage division, which is rapid and marks the start of cellular division. Though the timing of these early stages is similar across many species, some notable distinctions exist in the number of cleavage divisions and the pattern in which they occur, particularly in species with differing yolk content in their eggs (Table **1**).

Table 1. Species-specific differences in embryonic development.

Species	Fertilization (Day 0)	2-cell Stage (Day 1)	4-cell Stage (Day 2)	Early Morula (Day 3)	Late Morula (Day 4)	Blastocyst Formation (Day 5-6)	Implantation (Day 7-10)
Human	Fertilization, 1st cleavage	2-cell	4-cell	Early morula	Late morula	Day 5-6: Blastocyst formation	Days 7-10: Uterine wall implantation
Mouse	Fertilization, 1st cleavage	2-cell	4-cell	Early morula	Late morula	Day 5: Blastocyst formation	Day 6: Uterine wall implantation
Rat	Fertilization, 1st cleavage	2-cell	4-cell	Early morula	Late morula	Day 5-6: Blastocyst formation	Days 6-7: Uterine wall implantation
Frog	Fertilization, 1st cleavage	2-cell	4-cell	Early morula	Late morula	Day 5: Blastocyst-like formation	Not applicable (external fertilization)
Chicken (Hen)	Fertilization, 1st cleavage	2-cell	4-cell	Early morula	Late morula	Day 5: Blastoderm (similar to blastocyst)	Days 4-5: Implantation (yolk sac membrane)
Sheep (Ewe)	Fertilization, 1st cleavage	2-cell	4-cell	Early morula	Late morula	Day 6-7: Blastocyst formation	Days 7-8: Uterine wall implantation
Cow (Cow)	Fertilization, 1st cleavage	2-cell	4-cell	Early morula	Late morula	Day 7: Blastocyst formation	Days 7-9: Uterine wall implantation
Dog (Bitch)	Fertilization, 1st cleavage	2-cell	4-cell	Early morula	Late morula	Day 7-8: Blastocyst formation	Days 12-14: Uterine wall implantation
Cat (Queen)	Fertilization, 1st cleavage	2-cell	4-cell	Early morula	Late morula	Day 7-8: Blastocyst formation	Days 10-12: Uterine wall implantation

(Table 1) cont.....

Horse (Mare)	Fertilization, 1st cleavage	2-cell	4-cell	Early morula	Late morula	Day 7-8: Blastocyst formation	Days 8-9: Uterine wall implantation

- **Mammals (Human, Mouse, Rat, Sheep, Cow, Dog, Cat, Horse):** In mammals, fertilization typically leads to the rapid division of the zygote, first into two cells (Day 1), then into four cells (Day 2), and continuing through the stages of morula formation and blastocyst development. The fertilization process, followed by cleavage, progresses uniformly in species with minimal yolk content, ensuring similar early development.

- **Birds (Chicken) and Frog:** In contrast, species such as frogs and chickens exhibit notable variations in early cleavage patterns. The cleavage divisions in these species occur unevenly, with cells dividing more rapidly in areas with less yolk. Frogs, with moderate yolk content, undergo more localized cleavage, and chickens, with high yolk content, show even more pronounced variations.

2. Morula and Blastocyst Formation

After the early cleavage stages, the embryo reaches the morula stage, where a solid ball of cells forms. As cellular differentiation begins to occur, species-specific differences emerge in the timing and structure of these stages (Table **1**).

- **Mammals (Human, Mouse, Rat, Sheep, Cow, Dog, Cat, Horse):** In most mammals, the morula progresses to the blastocyst stage by Day 5 or 6. The blastocyst is characterized by the formation of a blastocoel (fluid-filled cavity), trophoblast, and inner cell mass (ICM), which plays a crucial role in the subsequent stages of implantation and development.

- **Frogs and Chickens:** While the term "blastocyst" does not apply to frogs and chickens, they undergo similar developmental events during the formation of the blastoderm, a structure that serves a similar purpose in these species. These embryos are still in the process of organizing their cellular structures into more defined layers.

3. Species-Specific Timing of Implantation

The timing of implantation, where the blastocyst embeds itself in the uterine wall, is another critical point of divergence between species (Table **1**).

- **Mammals (Human, Mouse, Rat, Sheep, Cow, Dog, Cat, Horse):** The timing of implantation varies slightly across different mammalian species. In humans and mice, implantation typically occurs between Days 7 and 10, while in other mammals like cows and horses, implantation can occur slightly later (Days 7-9). Dog and cat species show a delay in implantation compared to most other mammals, with implantation occurring between Days 10 and 14.

- **Birds and Amphibians:** Frogs and chickens do not undergo implantation as mammals do, since their embryos are deposited externally (in amphibians, in water or moist environments, and birds, in the egg). However, the developing embryo in birds relies on the yolk sac for nourishment and protection during early development, which is an essential parallel to implantation in mammals.

4. *Evolutionary and Developmental Implications*

The differences observed in the timing of cleavage, morula formation, and implantation reflect both evolutionary adaptations and the unique needs of each species (Table **1**). For example, species that lay eggs, such as birds and amphibians, have evolved distinct mechanisms to ensure that their embryos are nourished and protected until they are capable of independent life. In contrast, mammals rely on internal gestation, leading to variations in the implantation process that facilitate the exchange of nutrients between the mother and embryo.

Moreover, understanding these species-specific differences in early embryonic development helps researchers draw conclusions about the evolutionary processes that have shaped reproductive strategies. In mammals, internal development provides greater protection and more regulated growth conditions, while egg-laying species face different challenges related to external environmental factors, such as temperature and moisture.

These differences illustrate the fascinating diversity of reproductive strategies across species, providing valuable insights for both basic biology and applied fields such as agriculture, medicine, and conservation.

4. DEVELOPMENTAL MILESTONES: FROM GASTRULATION TO ORGANOGENESIS

4.1. Gastrulation: Orchestrating Embryonic Layers

Gastrulation marks a critical phase in embryonic development, facilitating the transformation of the blastula into a complex, three-dimensional organism characterized by distinct inner, middle, and outer layers. Upon completion, the **gastrula** comprises an outer **ectoderm**, an inner **endoderm**, and an intervening **mesoderm** layer, each poised to give rise to specific tissues and organs during subsequent developmental stages (Fig. **3**).

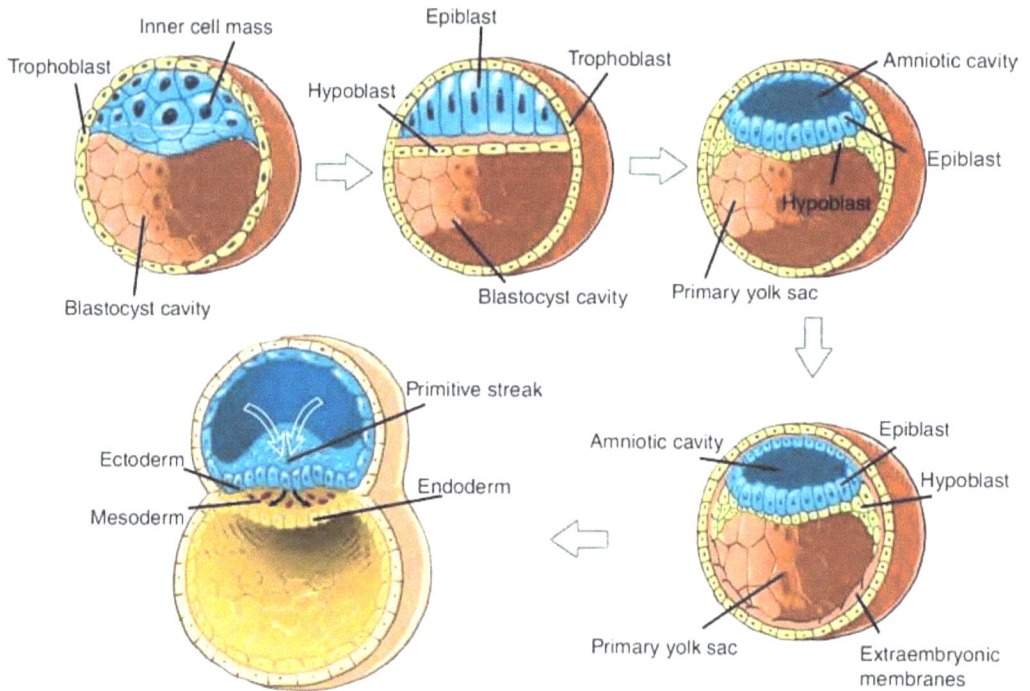

Fig. (3). Gastrulation, growth, and development.

A notable variation in gastrulation, involving cell migration towards endodermal and mesodermal positions *via* the thickened **primitive streak**, has been observed in reptiles and persists in avian and mammalian embryos. This evolutionary continuity lends credence to the hypothesis of birds and mammals evolving from reptilian ancestors, underscoring the intricate interplay between ontogeny and phylogeny.

4.2. Organogenesis: Sculpting Functional Tissues and Organs

Organogenesis, the hallmark of tissue and organ formation during embryonic development, entails the specialization of cells within the embryo and on its surface. Comprising **morphogenesis** and **differentiation**, organogenesis orchestrates intricate cellular changes that culminate in the emergence of diverse tissues and organs.

During morphogenesis, cells undergo dynamic shape alterations, exemplified by **neurulation** in vertebrate embryos, where the neural plate undergoes folding to initiate the formation of the brain and spinal cord. Concurrently, differentiation ensues, guiding cells towards maturation and functional specialization. This process encompasses the acquisition of function-specific morphologies, such as the elongated configuration of skeletal muscle cells.

Furthermore, cell differentiation endows cells with responsiveness, enabling regulation within the organism through hormonal, neuronal, and other signaling mechanisms. This multifaceted interplay between morphogenesis and differentiation underpins the orchestrated symphony of tissue and organ development, laying the foundation for organismal complexity and functionality.

5. PROTECTIVE SHELLS: EXTRAEMBRYONIC MEMBRANES IN LAND VERTEBRATES

During embryonic development in land vertebrates, the growing embryos are ensconced within four specialized extraembryonic membranes. These membranes serve as protective coverings while facilitating crucial exchanges of gases, nutrients, and other essential materials vital for embryonic growth and development.

6. GROWTH DYNAMICS: CELLULAR PROLIFERATION AND DEVELOPMENTAL PHASES

Embryonic growth predominantly stems from an amplification in cell numbers rather than an enlargement of individual cells. The availability of nutrients, particularly yolk in many species, often dictates the extent of embryonic growth. In organisms such as frogs and insects that undergo development independently of the maternal body, the transition to the larval stage enables self-feeding, followed by **metamorphosis** into the adult form.

In several species, the most pronounced growth occurs during the juvenile and adolescent phases, tapering off once the organism attains typical adult proportions. Although growth halts upon reaching adulthood, cellular turnover, including the replacement of dead cells, may persist.

Moreover, certain species exhibit regenerative abilities (**regeneration**), wherein adults can **regenerate** lost body parts. This process involves the **dedifferentiation** of stump tissue cells, leading to rapid division and subsequent generation of cells essential for regrowth. Additionally, **compensatory hypertrophy**, characterized by increased mass and cell numbers in residual tissues, offers a temporary response without dedifferentiation.

7. AGING DYNAMICS: THE CULMINATION OF DEVELOPMENT

Aging represents an ongoing developmental process marked by the gradual deterioration of bodily structures and functions over time. Proposed mechanisms underlying aging include collagen degeneration, cellular senescence due to finite division capacities, immune system decline, and the accumulation of aging pigments like lipofuscins. These factors collectively contribute to the culmination of the developmental journey, leading to eventual death.

Vocabulary (Index of Terms)

Acrosome reaction [ˈækrəˌsoʊm riˈækʃən]	The acrosome reaction refers to a series of observable structural changes undergone by a sperm cell when it is in the vicinity of an ovum in the oviduct. Specifically, during this process, numerous openings appear in the membrane of the sperm head, allowing the contents of the acrosome to be released. This reaction is a crucial step in fertilization, as it enables the sperm to penetrate the protective layers surrounding the ovum and ultimately fuse with it to initiate the formation of a zygote.
Blastomere [ˈblæs.təˌmɪər]	The ascus is a distinctive structure found in Ascomycetes, responsible for producing sexual spores. A blastomere is one of several small cells formed from an animal zygote during the process of cleavage. Cleavage is the rapid division of the zygote into smaller cells called blastomeres. Each blastomere contains a portion of the zygote's cytoplasm and genetic material. These cells continue to divide through successive rounds of cleavage, eventually forming a hollow ball of cells known as a blastula. Blastomeres play a fundamental role in embryonic development, contributing to the formation of various tissues and organs as the embryo grows and develops.
Blastula [ˈblæs.tjʊlə]	The blastula is the stage of the early animal embryo that follows cleavage and precedes gastrulation. Typically, a blastula is composed of a hollow ball of cells, with a wall consisting of one to several layers of cells. The central cavity of the blastula is known as the blastocoel. During blastulation, cells divide rapidly, forming a hollow sphere of cells that marks an essential stage in embryonic

development. The blastula provides the foundation for the next stage, gastrulation, during which the embryo undergoes significant morphogenetic movements to establish the three primary germ layers: ectoderm, mesoderm, and endoderm.

Cleavage [ˈkliːvɪdʒ]

Cleavage refers to the mitotic division of the zygote that occurs immediately after fertilization. During this process, the zygote undergoes rapid cell divisions, resulting in the formation of a ball of smaller cells called blastomeres. Importantly, cleavage occurs without an overall increase in the size of the embryo. Instead, the zygote divides into increasingly smaller cells, each containing a portion of the original genetic material. Cleavage is a critical stage in embryonic development, laying the groundwork for subsequent stages such as blastulation and gastrulation.

Compensatory hypertrophy [ˌkɒmpənˈseɪtəri haɪˈpɜːtrəfi]

Compensatory hypertrophy is a physiological process wherein an organ or tissue increases in size in response to the loss or dysfunction of a similar or opposing structure. This term is often associated with the concept of regeneration, particularly in the context of tissues or organs that have the capacity to regrow or compensate for damage. In compensatory hypertrophy, the remaining healthy tissue undergoes enlargement to compensate for the lost or impaired function of the damaged tissue. This phenomenon is observed in various biological systems, including skeletal muscle, liver, and kidney, and it plays a crucial role in maintaining homeostasis and functional integrity in response to an injury or a disease.

Dedifferentiation [diˌdɪfəˌrɛnʃiˈeɪʃən]

Dedifferentiation refers to the process in which specialized cells lose their specific characteristics and revert to a more primitive state. This phenomenon is often observed in regenerative processes, such as in the formation of a blastema at the site of a vertebrate limb stump.

Differentiation [ˌdɪfəˌrɛnʃiˈeɪʃən]

Differentiation refers to the complex process involving the progressive diversification of the structure and functioning of cells within an organism. As cells differentiate, they undergo changes that lead to the development of specific characteristics and functions suited to their roles in the organism. This process results in a continual restriction of the types of transcription that each cell can undertake, meaning that cells become specialized   for   particular   functions   through differentiation.

Ectoderm ['ɛk.tə ˌdɜrm]

The ectoderm is the outermost of the three primary germ layers found in metazoan embryos. During embryonic development, it gives rise to various structures, primarily including the epidermal tissue, the nervous system, sense organs, and in lower forms, the nephridia.

Endoderm ['ɛn.də ˌdɜrm]

The endoderm is the innermost of the three primary germ layers found in metazoan embryos. During embryonic development, it gives rise to the epithelial lining of the digestive tract and associated organs, as well as to certain other internal organs such as the liver, pancreas, and lungs.

Fertilization [ˌfɜːr.tɪ.laɪˈzeɪ.ʃən]

Fertilization is the process in sexual reproduction where the haploid nuclei from two gametes, typically one from an egg and one from a sperm cell, unite to form a diploid cell called a zygote. This fusion of genetic material results in the creation of a new organism with a unique combination of genetic traits inherited from both parents.

Gastrula ['gæs.trʊ.lə]

The gastrula is a stage in the development of animal embryos that follows the blastula and results from gastrulation. During gastrulation, cells of the blastula undergo   extensive   rearrangement,   leading   to   the formation of three primary germ layers: ectoderm, mesoderm, and endoderm. The gastrula is characterized by the differentiation of these germ layers, each destined to give rise to specific tissues and organs. Additionally, the central cavity of the gastrula opens to the exterior through a structure called the blastopore, which plays a crucial role in the organization of the developing embryo.

Gastrulation [ˌgæstrʊˈleɪʃən]

Gastrulation is the process in the animal embryo whereby a blastula undergoes a transformation into a gastrula, marked by the formation and organization of germ layers. During gastrulation, cells in the blastula undergo coordinated movements and rearrangements, leading to the establishment of three primary germ layers: ectoderm, mesoderm, and endoderm. This process is essential for the subsequent development of various tissues and organs in the organism.

Gene amplification [dʒiːn ˌæmplɪfɪˈkeɪʃən]

Gene amplification refers to a temporary and significant increase in the number of copies of a particular gene within a genome during a specific developmental period. This process leads to an elevated expression of the gene and can occur in response to various stimuli or developmental cues. Gene amplification plays a crucial role in regulating gene expression levels and can have important implications for cellular function and organismal development.

Mesoderm [ˈmɛs.ə ˌdɜrm]

The mesoderm is the middle layer of the three primary germ layers found in triploblastic animal embryos. During embryonic development, it gives rise to a diverse array of structures, including cartilage, bone, muscle, blood, kidneys, and gonads.

Metamorphosis [ˌmɛtəˈmɔːrfəsɪs]

Metamorphosis refers to the process of transformation from the larval to the adult form in certain organisms, typically observed in amphibians, insects, and some other invertebrates.

Morphogenesis [ˌmɔːrfəˈdʒɛnɪsɪs]

Morphogenesis encompasses the developmental processes leading to the characteristic mature form of an organism or a specific part of an organism. This includes the shaping of tissues, organs, and overall body structures during embryonic development and growth.

Neurulation [ˌnjʊərʊˈleɪʃən]

Neurulation is the developmental stage in an embryo during which cells differentiate into the brain and spinal cord. This process involves the formation of the neural tube from the neural plate, which subsequently gives rise to the central nervous system.

Oogenesis [ˌoʊəˈdʒɛnɪsɪs]

Oogenesis is the specific term used to describe the process of gametogenesis that leads to the formation of eggs (ova) in female organisms. This process involves the maturation and development of oocytes within ovaries.

Organogenesis [ˌɔːrɡənoʊˈdʒɛnɪsɪs]

Organogenesis refers to the process of formation and development of organs within an organism. This complex process involves the differentiation, growth, and patterning of cells to create the diverse structures that make up the various organs of the body.

Ovaries [ˈoʊvəriz]

Ovaries are the female reproductive organs responsible for producing haploid sex cells, known as eggs or ova. These organs also secrete hormones such as estrogen and progesterone, which regulate the menstrual cycle and other aspects of female reproductive physiology.

Oviduct [ˈoʊvɪdʌkt]

The oviduct, also known as the fallopian tube, is a tube-like structure in the female reproductive system that carries the primary oocyte from the ovary to the uterus. It is the site where fertilization typically occurs when a sperm cell meets the egg as it travels through the oviduct.

Ovum [ˈoʊvəm] / Ova [ˈoʊvə]

An ovum, or ova in plural, refers to an unfertilized, non-motile female gamete. In many animals, including humans, ova are produced within the ovaries and are released during ovulation. They serve as the female counterpart to sperm cells and are essential for sexual reproduction.

Parthenogenesis [ˌpɑːrθənoʊˈdʒɛnɪsɪs]

Parthenogenesis is a form of reproduction in which eggs develop normally without being fertilized by a male gamete. This process results in the production of an individual usually genetically identical to the parent. Parthenogenesis occurs in various organisms, including some insects, reptiles, and certain plants.

Primitive streak [ˈprɪmətɪv striːk]

The primitive streak is a longitudinal band of embryonic mesodermal cells that develops within a dorsal groove along the length of the gastrula in mammals and birds. It plays a crucial role in organizing the body plan of the developing embryo and is involved in the process of gastrulation, during which the three primary germ layers (ectoderm, mesoderm, and endoderm) are established.

Regeneration [rɪˌdʒɛnəˈreɪʃən]

Regeneration is the process by which an organism replaces tissues or organs that have been lost or damaged. This phenomenon is observed in various organisms across the animal kingdom, allowing them to restore lost body parts and maintain functionality.

Sperm [spɜːrm] / sperm [spɜːrm]

Sperm is the singular and plural form of the word. It refers to the male reproductive cells that develop from spermatids by losing much of their cytoplasm and developing long tails. Sperm cells are the male gametes responsible for fertilizing the female egg during sexual reproduction.

Spermatogenesis [ˌspɜːrmətoʊˈdʒɛnəsɪs]

Spermatogenesis is the specific name given to the process of gametogenesis that leads to the formation of sperm in the testes of male organisms. During spermatogenesis, spermatogonia undergo a series of mitotic and meiotic divisions, ultimately producing mature sperm cells (spermatozoa) capable of fertilizing a female egg.

Yolk [joʊk]

Yolk is the store of food material, consisting mostly of protein and fat, that is present in the eggs of most animals. It provides nourishment to the developing embryo during its early stages of development.

Zygote [ˈzaɪgoʊt]

A zygote is a diploid cell that results from the union of an egg and a sperm during fertilization. It marks the beginning of the development of a new organism and contains the combined genetic material from both parents.

COMPREHENSION TASKS

I. Key Terms: Matching

Instructions: Match each term from the left column with the corresponding description from the right column. Write the letter of the correct description next to the number of the term. Each description will only match with one term. Good luck!

1.Yolk	A.Gamete or egg
2.Amnion	B.Spermatogonia
3.Regeneration	C.Homologous with ovaries
4.Parthenogenesis	D.Egg tube
5.Cleavage	E.Prevents multiple fertilizations
6.Zygote	F."virgin" birth
7.Chorion	G.Individual blastula cells
8.Testis	H.Divides a single-celled zygote into many small cells
9.Cortical reaction·	I.Food
10.Ovum	J.Fertilized egg
11.Primitive streak	K.Associated with large yolks
12.Blastomere	L.Cushions embryo
13.Allantois	M.M.'trash dump
14.Oviduct	N.Fuses with allantois
15.Gonial cell	O.Replacement of lost parts

Cell Biology True or False Quiz: Determine whether the following statements are true or false based on your understanding of cell biology concepts. Write "True" if the statement is correct and "False" if the statement is incorrect. Good luck!

1. ----- Sertoli cells play a supportive role in sperm production.

2. ----- The sperm head contains a limited quantity of mitochondria.

3. ----- In certain species, oogenesis can halt for extended periods.

4. ----- An ostrich egg may be regarded as a singular cell entity.

5. ----- During the "lampbrush" phase, chromosomes generate minimal mRNA.

6. ----- The egg's jelly coat triggers the acrosome reaction.

7. ----- Before fertilization, the egg carries a positive electrical charge.

8. ----- Parthenogenesis results exclusively in the birth of females.

9. ----- The so-called "vegetable pole" harbors the highest concentration of yolk.

10. ----- The gut cavity originates from an archenteron.

11. ----- Amphioxus eggs exhibit a distinct primitive streak.

II. **Completion Exercise: Complete each sentence by filling in the blanks with the appropriate term or phrase**. Write your answers in the space provided. Good luck!

1. The heads of spermatids are enveloped by _____ cells.

2. The sperm acrosome originates from the _____.

3. Another term for "egg white" is _____.

4. The outermost protective layer of the unfertilized egg is known as the _____.

5. The final stage of the cortical reaction involves the elevation of the _____.

6. Following fertilization, the mammalian zygote divides to form the _____.

7. The alteration in shape of cells and cell populations is termed as _____.

8. The neural tube begins to form during the process of _____.

III. **Multiple Choice:** Choose the best answer for each question by circling the corresponding letter.

1. Sperm are produced and matured in the _____.

a. Acrosome

b. Supporting Sertoli cells

c. Seminiferous tubules

d. Vas deferens

e. Follicle

2. The energy required to power the sperm flagellum is produced by _____.

a. Mitochondria

b. The acrosome

c. Glucose

d. Glycogen

e. None of the above

3. Meiosis concludes in egg cells _____.

a. Within the ovary

b. At ovulation

c. Within the oocyte

d. Just before fertilization

e. Just after fertilization

4. Nurse cells _____.

a. Merge with sperm during fertilization

b. Give rise to mature ova

c. Synthesize proteins and nucleic acids crucial for embryonic development

d. Provide nutrients for mature sperm

e. Enhance fertilization

5. The most crucial factor in determining whether eggs and sperm unite is _____.

a. Vaginal pH

b. Sperm mobility

c. Sperm count

d. Mating behavior

e. Timing of the egg's release

6. The entry of multiple sperms into a mature egg is hindered by _____.

a. Barriers produced by sperm

b. Reaction of the egg's outer layer

c. Decrease in internal pH

d. Degradation of the membrane surrounding the egg

e. All of the above

7. The spontaneous progression of an unfertilized egg is termed as _____.

a. Differentiation

b. Gametogenesis

c. Oogenesis

d. Parthenogenesis

e. Morphogenesis

8. The cleavage patterns in the zygote are shaped by _____.

a. Cytoplasm

b. Yolk

c. DNA

d. Sperm

e. The nucleus

9. The result of cell cleavage is a _____.

a. Blastocoele

b. Blastomere

c. Blastula

d. Gastrula

e. Blastopore

10. Studies on sea squirts have shown that by the end of cell division _____.

a. Cells are randomly distributed throughout the embryo

b. Cells are organized according to their future function

c. Gray crescent cells are randomly distributed in the embryo

d. Yellow crescent cells give rise to the primary skeletal organs

e.   Three distinct layers are visible in the embryo

11. The zygote advances to a three-tiered phase during _____.

a. Amplification

b. Gastrulation

c. Morulation

d. Blastulation

e. Cleavage

12. In birds, gastrulation occurs through cellular migration across the _____.

a. Blastopore

b. Ectoderm

c. Primitive streak

d. Mesoderm

e. Blastodisk

13. The ectoderm forms into _____.

a. Muscle

b. Skin

c. The heart

d. The lungs

e. All of the above

14. A cell that can exclusively generate cells with a particular function is termed as _____.

a. A germ cell

b. Differentiated

c. Fixed

d. Diploid

e. Altered

15. Morphogenesis entails an alteration in _____.

a. Cell function

b. DNA activity

c. Cell shape

d. RNA activity

e. None of the above

16. The neural tube forms _____.

a. By neurulation

b. By upward folding of the neural plate

c. From the ectodermal layer

d. All of the above

e. By upward folding and differentiation of the mesoderm

17. The maximum size attainable by a chicken embryo is restricted by _____.

a. Protein synthesis

b. The size of its cells

c. Cell differentiation

d. The amount of food in the egg

e. Its age

18. The development of cellular sensitivity is linked with _____.

a. Gastrulation

b. Adult growth

c. Regeneration

d. Morphogenesis

e. Differentiation

19. One fundamental disparity between dedifferentiation and compensatory hypertrophy is that _____.

a. Mitosis exclusively transpires during dedifferentiation

b. Regeneration is exclusive to amphibians

c. Dedifferentiation solely transpires during regeneration

d. Regeneration transpires in humans while compensatory hypertrophy occurs in amphibians

e. Dedifferentiation exclusively transpires during compensatory hypertrophy

20. All of the following theories have been suggested as potential causes of aging except_____.

a. Accumulation of lipofuscin

b. Breakdown of collagen

c. Constraints on cell division count

d. Dedifferentiation

e. Weakening of the immune system

READING UNDERSTANDING

Diverse Feeding Strategies Among Warbler Species in Spruce-Fir Forests

In closely related groups, variations in feeding techniques can be observed. Within the spruce-fir forests of northern New England and Canada, five warbler species coexist peacefully in the same trees, each exhibiting slightly different feeding behaviors. For instance, the Cape May warbler predominantly seeks insects in the uppermost parts of trees, towards the outer edges of branches. Conversely, the yellow-rumped warbler forages closer to the tree trunk, focusing on lower branches and even the ground. These two species also engage in aerial insect capture by traversing between trees. The black-throated green warbler primarily feeds at mid-level elevations, often on branch tips, and employs hovering to access insects more frequently than other species. Similarly, the Blackburnian warbler targets the outermost tips of branches but at a slightly lower height. In contrast, the bay-breasted warbler concentrates its feeding efforts in the lower half of the trees, often avoiding the outermost branches. Through such niche partitioning, these five species effectively exploit what might appear to be a single food source within the tree canopy.

1. What is the primary objective of the author in this excerpt?

A. Encouraging harmonious coexistence among groups

B. Enumerating the varieties of insects consumed by warblers

C. Demonstrating how varied feeding behaviors facilitate species cohabitation

D. Providing guidance to naturalists regarding the locations of different warbler species

2. Based on the passage, it can be deduced that warblers are ___.

A. Cats

B. Birds

C. Insects

D. Squirrels

3. As per the passage, which of the following warblers is likely to hover more than the others?

A. The Cape May warbler

B. The yellow-rumped warbler

C. The bay-breasted warbler

D. The black-throated green warbler

4. Where should an individual aiming to spot a Blackburnian warbler direct their attention?

A. On the ground

B. At the base of the tree

C. At the tips of the lower branches

D. At the tips of the upper branches

5. The five warbler species can utilize a single food source by __.

A. Specializing in different types of insects

B. Inhabiting various regions of the forest

C. Consuming different segments of the trees

D. Scouring for food in distinct parts of the trees

6. As used throughout the passage, the term "trunk" pertains to the __.

A. Central part of a tree

B. Primary part of an animal's body

C. Flexible snout of an animal

D. Luggage compartment of a car

7. The excerpt is most likely to be featured in a textbook concerning ___.

A. Botany

B. Zoology

C. Geography

D. Agriculture

READING MATERIALS

Animals

1. Exploring the Complexity of Animal Morphology and Embryonic Development

Animals exhibit remarkable diversity in their morphology, ranging from diminutive forms smaller than many protists to colossal entities such as whales and giant squid. Cellular structures within animals display a vast array of forms and functions, constituting the intricate compositions of their bodies, which function adeptly across a broad spectrum of environments. Unlike sponges, animal cells are organized into tissues, cohesive assemblies of cells serving as structural and functional units. Typically, these tissues further integrate into organs, intricate assemblies comprising two or more types of tissues.

Sexual reproduction predominates among animals, where gametes—sperm and eggs—undergo no mitotic division. With rare exceptions, animals maintain a diploid state throughout their lifecycle, with gametes representing the sole haploid cells. The intricate body plan of an animal arises from the fusion of male and female gametes to form a zygote. This zygote undergoes a characteristic process of embryonic development, initially progressing through a series of mitotic divisions to form a hollow sphere of cells known as the blastula, a developmental stage universally observed across animals.

In most animals, the blastula undergoes invagination at a specific site, resulting in the formation of a hollow sac with an aperture at one end termed the blastopore, thus transforming into a gastrula. Subsequent cellular growth and migration within the gastrula contribute to the formation of the digestive system. Variations in the early stages of embryonic development are discernible among animal phyla, offering crucial insights into their evolutionary interrelations.

2. Unveiling the Enigmatic World of Sponges: Primitive Creatures Defying Traditional Animal Classification

Exploring the Unique Anatomy, Physiology, and Evolutionary Origins of Sponges: Sponges, the most basic of all animals, lack organized tissues and organs, presenting a stark contrast to more complex organisms. Their bodies, comprised primarily of loosely aggregated cells ensconced within a gel-like substance, exhibit minimal cellular coordination, allowing individual clusters to disperse and reform even through fine meshes.

In their juvenile stage, sponges assume a sac or vase-like shape, with an outer layer of flattened cells known as the epithelial wall encasing specialized flagellated cells called choanocytes or collar cells. Embedded within a protein-rich matrix between the choanocytes and epithelial wall are various cellular components, including amoeboid cells, calcium carbonate or silica spicules, and spongin fibers, contributing to both structural reinforcement and predator deterrence.

Perforated by numerous small pores, the sponge's body facilitates the ingress and egress of water, a function critical for feeding, respiration, and waste elimination. The coordinated beating of flagella on choanocytes propels water through the sponge's internal channels, a primitive circulatory mechanism reminiscent of more complex animals.

Examination at the microscopic level reveals intricate substructures within choanocytes, notably the collar of hair-like projections surrounding the base of each flagellum, facilitating the capture and ingestion of food particles suspended in the water. The resemblance between choanocytes and certain unicellular organisms, such as choanoflagellates, suggests a probable evolutionary ancestry, with choanoflagellates likely serving as precursors to sponges and potentially other animal groups.

Sponges exhibit remarkable diversity, comprising approximately 10,000 marine species alongside an additional 150 freshwater species, spanning a wide range of sizes and inhabiting various depths in aquatic environments. While some species remain diminutive, others attain substantial proportions, such as the loggerhead sponges, reaching diameters exceeding 2 meters. Although larval stages of sponges are mobile, adults typically adopt a sessile lifestyle, firmly anchored in place within their habitat.

3. Diversity of the Arthropods

Arthropods, a diverse phylum within the animal kingdom, exhibit a wide array of mouthpart adaptations. Among them, mandibulates and chelicerates represent two primary categories, each distinguished by unique anatomical features and evolutionary histories.

Many arthropods possess mandibles, specialized jaws formed through the modification of one pair of anterior appendages, excluding the closest to the anterior end. This group, known as mandibulates, encompasses crustaceans, insects, centipedes, millipedes, and a few other minor taxa. In contrast, chelicerates, comprising spiders, mites, scorpions, and related species, lack mandibles. Instead, they possess mouthparts called chelicerae, evolved from the appendages nearest the anterior end. These chelicerae often adopt the form of pincers or fangs, serving various feeding and defensive functions.

Interestingly, crustaceans appear to have independently evolved mandibles, distinct from those of insects and other mandibulates. This suggests separate evolutionary pathways for mandible development in crustaceans, further highlighting the intricate evolutionary relationships within the arthropod phylum.

3.1. Chelicerates: Exploring Ancient Lineages

The chelicerates, an ancient and diverse group within the arthropod phylum, boast a fossil record dating back approximately 630 million years, making them among the earliest multicellular animals documented in the geological record. Notably, the trilobites, a major group of extinct arthropods, were abundant during this era, with living horseshoe crabs believed to be direct descendants.

Among chelicerates, the class Arachnida reigns supreme as the largest and most diversified, encompassing approximately 57,000 named species. This predominantly terrestrial class includes a myriad of taxa such as spiders, ticks, mites, scorpions, and daddy longlegs. Scorpions, regarded as one of the most ancient groups of terrestrial arthropods, have been traced back to the Silurian Period, spanning approximately 425 million years.

Of particular note within the class, Arachnida are spiders (order Araneae), boasting around 35,000 named species globally. Additionally, the order Acarai, comprising mites, stands out as the largest and most diverse among the arachnids in terms of species richness. These minute organisms play pivotal roles in various ecosystems, contributing to ecological processes and biodiversity on a global scale.

3.2. Mandibulates: Exploring the World of Crustaceans

Mandibulates, specifically crustaceans belonging to the subphylum Crustacea, constitute a vast and diverse group of predominantly aquatic organisms, comprising approximately 35,000 species. This eclectic assemblage includes familiar inhabitants of marine and freshwater ecosystems, such as crabs, shrimps, lobsters, crayfish, barnacles, water fleas, and pillbugs, among others.

Crustaceans play indispensable roles in aquatic ecosystems, often being highly abundant and exerting significant influence on ecosystem dynamics. Given their ubiquity and ecological importance, crustaceans have earned the moniker "the insects of the water," underscoring their ecological significance.

Most crustaceans exhibit distinctive features such as two pairs of antennae, three pairs of chewing appendages, and varying numbers of pairs of legs. Notably, crustaceans differ from insects but share similarities with centipedes and millipedes in possessing legs on both their abdomen and thorax. A unique characteristic distinguishing crustaceans from other arthropods is their possession of two pairs of antennae, a feature exclusive to this taxonomic group.

Through their diverse adaptations and ecological roles, crustaceans contribute to the intricate web of life in aquatic ecosystems, serving as vital components of food webs and playing crucial roles in nutrient cycling and ecosystem functioning.

3.3. Mandibulates: Unraveling the World of Insects

Insects, classified under the class Insecta, stand as the most extensive group of arthropods, boasting unparalleled diversity both in terms of species numbers and individual populations. Their omnipresence across virtually all terrestrial and freshwater habitats, and even certain marine environments, underscores their status as the most abundant eukaryotic organisms on the planet. Remarkably, it has been estimated that for every human inhabitant, there are approximately 300 million insects alive at any given moment, illustrating their astonishing abundance.

Comprising more than 70% of all named animal species, insects play pivotal roles in ecosystems worldwide, with millions of additional species awaiting discovery and classification. In the United States and Canada alone, approximately 90,000 insect species have been described, with the actual number likely surpassing 125,000.

Structurally, insects exhibit a distinctive body plan characterized by three main sections: the head, thorax, and abdomen. They possess three pairs of legs attached to the thorax, along with one pair of antennae. Most insects feature compound eyes composed of numerous independent visual units, facilitating diverse visual capabilities adapted to their ecological niches.

The thorax of insects, comprising three segments, predominantly houses muscles responsible for operating both the legs and wings. While insects typically possess two pairs of wings attached to thoracic segments, evolutionary adaptations have led to the loss of wing pairs in certain groups such as flies. The wings of adult insects are primarily composed of chitin, providing structural support and protection during flight.

In terms of respiration, insects lack a single major respiratory organ, instead relying on a network of small, branched air ducts known as tracheae. These tubes, lined with cuticle, facilitate the transmission of oxygen throughout the insect's body, ultimately branching into minute tracheoles. Air enters the tracheal system through specialized openings called spiracles, distributed across the thorax and abdomen segments.

Characteristic of insects is their diverse array of metamorphic patterns, wherein distinct life stages succeed one another. While some insects undergo gradual metamorphosis, with stages transitioning smoothly, others experience abrupt changes between life stages, exemplifying the remarkable plasticity and adaptability inherent in insect biology.

4. Chordates: Unveiling the Essence of Vertebrate Evolution

Chordates, comprising the phylum Chordata, stand as one of the most extensively studied and recognizable groups within the animal kingdom. With approximately 42,500 species, including birds, reptiles, amphibians, fishes, and mammals, chordates encompass a diverse array of organisms, including our own species.

Key to the chordate identity are three defining features: a single hollow nerve cord, a notochord, and pharyngeal slits. These structures have played pivotal roles in the evolutionary trajectory of the phylum. In more advanced vertebrates, the dorsal nerve cord undergoes differentiation into the brain and spinal cord. The notochord, persisting throughout the life cycle of some invertebrate chordates, becomes enveloped and eventually replaced by the vertebral column during embryological development in vertebrates. While pharyngeal slits are present in all vertebrate embryos, they are typically lost during the development of terrestrial vertebrates, serving as a vestige of their aquatic ancestry.

Beyond these foundational features, chordates possess additional characteristics that distinguish them from other animal groups. Their body plan often exhibits segmentation, with distinct blocks of muscles visible in less specialized forms. The presence of an internal skeleton or notochord facilitates powerful locomotion, a hallmark of chordates. Furthermore, chordates typically possess a tail extending beyond the anus during embryonic development, a feature uncommon among other animals which typically exhibit a terminal anus.

Within the vast phylum of chordates, three major groups emerge. The acraniates, consisting of tunicates and lancelets, lack a brain. In contrast, the craniate chordates, encompassing vertebrates, exhibit well-defined cranial structures. While further exploration will delve into the intricacies of these chordate subphyla, the vertebrates stand as a testament to the evolutionary innovations and adaptations that have shaped the chordate lineage.

4.1. Tunicates: Delving into the Primitive Ancestry of Chordates

Tunicates, belonging to the subphylum Tunicata, represent a group of approximately 1,250 marine species, primarily characterized by their sessile nature as adults. Despite their stationary lifestyle, adult tunicates lack discernible signs of segmentation or a defined body cavity, rendering the identification of evolutionary relationships challenging based solely on their adult features.

However, the tadpole-like larvae of tunicates offer critical insights into their evolutionary heritage, as they conspicuously exhibit fundamental chordate characteristics such as a notochord and nerve cord. These larvae, though rudimentary in their development, serve as a testament to the primitive combination of features observed in tunicates. Despite their chordate traits, tunicate larvae display limited feeding capabilities and possess underdeveloped digestive systems. After a brief period of free-swimming, the larvae settle and attach themselves to suitable substrates using a sucker-like structure.

4.2. Lancelets: Exploring the Fishlike Marvels of Cephalochordates

Lancelets, categorized under the subphylum Cephalochordata, are scaleless marine chordates resembling small fish, typically a few centimeters in length. Widely distributed in shallow waters across the world's oceans, lancelets derive their English name from their resemblance to a lancet, a small surgical knife.

In contrast to tunicates, lancelets retain a notochord throughout their entire lifespan, extending along the length of the dorsal nerve cord. Their fishlike appearance belies their primitive chordate characteristics, highlighting their significance in understanding the evolutionary transitions within the chordate lineage.

4.3. Vertebrates: Embarking on the Evolutionary Journey of Craniate Chordates

Vertebrates, constituting the subphylum Vertebrata, distinguish themselves from other chordates by the presence of a vertebral column, which typically replaces the notochord in adult individuals to varying degrees. Additionally, vertebrates, also known as craniate chordates, possess a distinct head housing a skull and brain, indicative of their advanced neurological development.

The hollow dorsal nerve cord of most vertebrates finds protection within a U-shaped groove formed by paired projections from the vertebral column, underscoring the intricate adaptations characteristic of this group. Vertebrates encompass diverse classes, including Agnatha (lampreys and hagfishes), Chondrichthyes (cartilaginous fishes such as sharks and rays), and Osteichthyes (bony fishes, the predominant fish group today).

Furthermore, vertebrates extend beyond aquatic realms to include tetrapods, which encompass Amphibia (amphibians), Reptilia (reptiles), Aves (birds), and Mammalia (mammals), each representing distinct evolutionary branches within the vertebrate lineage.

Biology for Students, 2025, 177-199

CHAPTER 8

Evolutionary Forces Shaping Species Diversity

Abstract: Understanding species diversity requires an exploration of various evolutionary forces and mechanisms. Modern biologists define a species as a group capable of interbreeding to produce viable offspring, though this concept primarily applies to sexually reproducing organisms. Asexual organisms are classified based on physical traits. Mechanisms such as prezygotic and postzygotic isolation prevent gene exchange between closely related groups. Prezygotic mechanisms include ecological, behavioral, mechanical, and temporal isolations, while postzygotic mechanisms result in hybrid sterility or breakdown. Ernst Mayr's allopatric speciation model outlines how geographic barriers and subsequent genetic divergence lead to speciation. Genetic identity measures the proportion of shared structural genes, and processes like polyploidization can drive rapid divergence. Macroevolution encompasses large-scale transformations, such as divergent and convergent evolution, and is often inferred through phylogenetic analysis and the fossil record. Microevolutionary processes contribute to macroevolutionary patterns, with ongoing research investigating the mechanisms underlying significant evolutionary changes.

Keywords: Allopatric speciation, Gene exchange, Macroevolution, Speciation.

UNDERSTANDING THE CHAPTER'S FOCUS: A ROADMAP

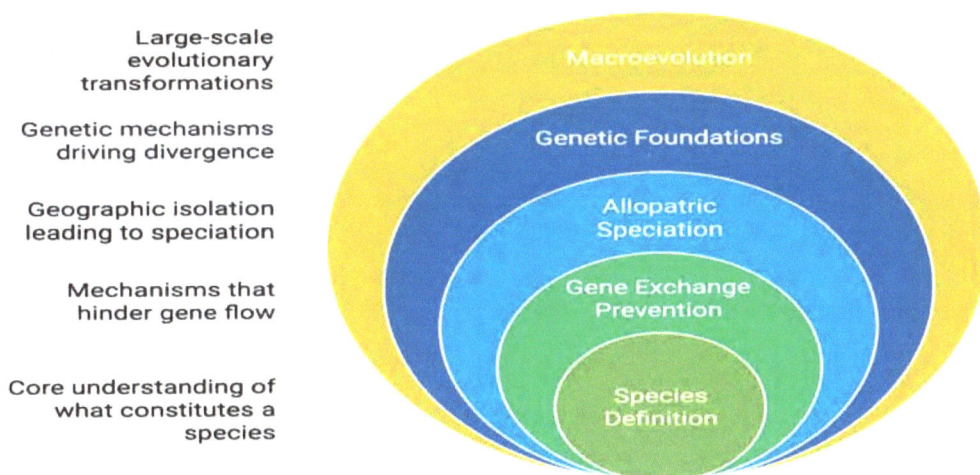

Evolutionary Forces Shaping Species Diversity

Mohammad Mehdi Ommati

1. HOW BIOLOGISTS DEFINE A SPECIES

Modern biologists typically define a species as a group of populations that can interbreed or have the potential to do so, maintaining reproductive isolation from other such groups. This means that individuals within a species can mate and produce viable offspring with one another, but cannot successfully reproduce with members of different species. This reproductive isolation concept is particularly precise, though it only applies to sexually reproducing organisms. For asexual reproducers, such as many prokaryotes, various plants, and certain animals, species classification relies on observable physical traits, including biochemical and morphological characteristics.

2. MECHANISMS OF GENE EXCHANGE PREVENTION

Two primary mechanisms function to impede gene exchange among closely related groups. The first mechanism comprises prezygotic isolating mechanisms, which hinder zygote formation. Prezygotic isolation encompasses ecological and behavioral categories. Ecological isolation arises when related groups adapt to slightly different environments over time, leading to genetic distinctions that hinder successful cross-fertilization. In behavioral isolation, distinct behaviors evolve among related groups, such as unique mating rituals, limiting gene exchange within each group. Occasionally, prezygotic isolation stems from mechanical barriers, rendering mating physically impossible due to structural genital incompatibility or failure of sperm and egg surface molecules to bind. Temporal isolation represents another prezygotic mechanism, where time-related environmental cues triggering reproductive processes differ among related species.

Postzygotic isolating mechanisms permit mating but yield inviable or sterile hybrid offspring. A specific instance, **hybrid sterility**, includes hybrid breakdown, wherein successive generations following a cross display reduced reproductive success. This contrasts sharply with the outcome of crossbreeding between genetically distant members of the same species, often resulting in heterozygote advantage, known as hybrid vigor.

Populations of a species distributed across a wide geographical range often exhibit a **cline**—a gradual variation in one or more characteristics as each population adapts to its local environment. Along a cline, distinct subspecies may emerge, with individuals at either end often experiencing reproductive isolation.

3. EVOLUTION INTO SPECIES: ESTABLISHING GENETIC ISOLATION

Ernst Mayr's **allopatric speciation** model posits a two-stage process for species formation. Initially, populations of a common species become divided by a physical or geographical barrier. Consequently, genetic disparities emerge over time, leading to either pre- or postzygotic isolation between the separated groups. Subsequently, in the second stage, these divergent populations may reestablish contact. Should this occur, speciation reaches completion through the influence of natural selection.

4. GENETIC FOUNDATIONS OF SPECIATION

The degree of variation between populations undergoing divergence into distinct species, or among species that have already diverged, is quantified by a measure known as genetic identity. This metric reflects the relative proportion of shared structural genes among individuals within the compared groups. Generally, biologists posit that the genetic transitions leading to speciation occur gradually. Once a new species emerges, it typically exhibits a faster rate of genetic divergence from related species. In certain cases, such as within the primate order, significant differences in physical characteristics do not align with corresponding variations in structural genes. Consequently, scientists propose that minor alterations in regulatory genes may underlie many of the substantial changes driving speciation and the emergence of higher taxonomic groups.

One mechanism capable of rapidly driving genetic divergence among populations is polyploidization—a sudden increase in the entire set of chromosomes. This phenomenon can lead to **sympatric speciation**, where new species arise without geographical isolation. A process akin to polyploidization, involving chromosome rearrangements, has been suggested to elucidate the evolutionary origins of giant pandas. Evidently, species can originate through diverse mechanisms.

5. UNDERSTANDING MACROEVOLUTION: MAJOR TRANSFORMA- TIONS

The transformations responsible for species divergence are often termed **microevolution**, while those leading to significant phenotypic distinctions separating genera, classes, orders, and beyond are termed **macroevolution**. Some lineages can be traced through the fossil record, while others necessitate inference through comparisons among related extant organisms. Constructing lineages of descent over evolutionary time yields a **phylogeny**.

The rationale behind phylogeny construction is straightforward: it posits that similarities in body structure, biochemistry, reproductive strategies, and other organismal features can be utilized to trace common ancestry. However, this process is intricate due to the varied patterns of evolution. In cases of **parallel evolution**, multiple lineages evolve in similar directions, while **convergent evolution** sees distantly related lineages converging due to analogous environmental pressures. Consequently, similar structures in different organisms may signify **homology** (shared ancestry) or **analogy** (independent emergence of structures serving similar functions) (Fig. **1**).

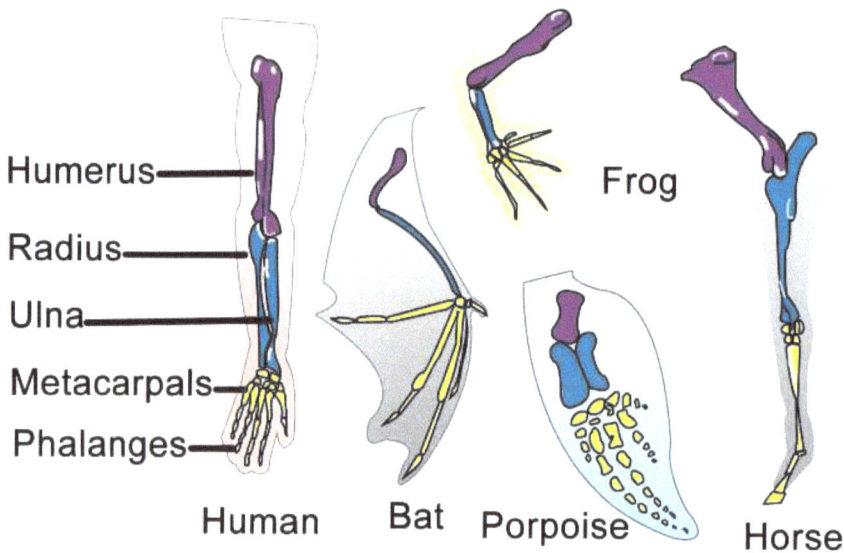

Fig. (1). Examples of homologous structures that reveal our shared ancestry.

One prevalent evolutionary pattern, evident in the fossil record, is **divergent evolution** or radiation, characterized by the branching and rebranching of a single lineage. **Extinction**, the complete loss of a species or group, is another common feature, with mass extinctions recorded at least five times in Earth's history.

Gaps in the fossil record have prompted some paleontologists to propose the **punctuated equilibrium** theory of evolution. This theory suggests that evolution occurs in bursts—marked by radical changes over short geological time periods—interspersed with periods of equilibrium. However, the theory remains controversial and tends to lack support when ample fossil evidence is available.

6. MICROEVOLUTION'S CONTRIBUTION TO MACROEVOLUTION

Many questions about large-scale evolutionary changes remain unanswered in biology. These inquiries range from whether novel higher taxa arise from yet-undescribed radical genetic mechanisms to the plausibility of known processes, such as genetic drift and minor genetic changes, accounting for the evolution of new genera, families, and orders. Researchers are investigating these domains using both traditional methods and the newer techniques of molecular biology.

Vocabulary (Index of Terms)

Allopatric speciation [ˌælə'pætrɪk ˌspiː'ʃi'eɪʃən]	Allopatric speciation refers to the emergence of separate species resulting from the differentiation of populations in geographic isolation.
Analogy [ə'nælədʒi]	Analogy refers to a similarity in function, and frequently in appearance, between two structures, resulting from convergent evolution rather than shared ancestry. This concept contrasts with homology.
Cline [klaɪn]	A cline refers to a gradual variation in characteristics across the range of a species, typically influenced by gradients of climate, soil, or other environmental factors.
Convergent evolution [kən'vɜːrdʒənt ˌiːvə'luːʃən]	Convergent evolution is an evolutionary phenomenon wherein distinct organisms exhibit similar characteristics.
Divergent evolution [daɪ'vɜːrdʒənt ˌiːvə'luːʃən]	Divergent evolution is a fundamental evolutionary pattern wherein individual speciation events result in the proliferation of branches within the evolutionary trajectory of a group of organisms.
Extinction [ɪk'stɪŋkʃən]	Extinction refers to the disappearance or cessation of a species, usually due to environmental factors or catastrophic events.
Homology [hə'mɒlədʒi]	Homology refers to a similarity between two structures resulting from inheritance from a common ancestor. The structures are described as homologous. This concept contrasts with analogy.
Hybrid sterility ['haɪbrɪd stə'rɪləti]	Hybrid sterility refers to the inability of hybrids between different species to produce viable offspring.

Macroevolution
[ˌmækroʊˌɛvəˈluːʃən]

Macroevolution refers to the broad-scale patterns, trends, and rates of change observed among groups of species over extended periods of time.

Microevolution
[ˌmaɪkroʊˌɛvəˈluːʃən]

Microevolution refers to changes in allele frequencies within a population, typically driven by processes such as mutation, genetic drift, gene flow, and natural selection.

Parallel evolution [ˈpærəˌlɛl ˌiːvəˈluːʃən]

Parallel evolution is the process by which organisms that were initially very similar evolve in the same direction, often resulting in comparable traits despite their separate lineages. This concept contrasts with convergent evolution.

Punctuated equilibrium
[ˈpʌŋktʃueɪtɪd ˌiːkwɪˈlɪbriəm]

Punctuated equilibrium is an evolutionary theory proposing that morphological changes occur rapidly within specific periods of time. During speciation, these changes take place in small populations, resulting in new species that are distinct from their ancestral forms. After speciation, species maintain a relatively stable form until extinction. This theory is distinct from the concept of phyletic gradualism.

Sympatric speciation [sɪmˈpætrɪk ˌspiːʃiˈeɪʃən]

Sympatric speciation occurs when new species evolve within the same geographical area, typically due to ecological, behavioral, or genetic barriers that arise within a single population. This process can occur suddenly, such as when polyploidy arises in a type of flowering plant capable of self-fertilization or asexual reproduction.

COMPREHENSION TASKS

I. Key Terms: Matching

Instructions: Match each term from the left column with the corresponding description from the right column. Write the letter of the correct description next to the number of the term. Each description will only match with one term. Good luck!

1. Analogy	A. Reproductively isolated
2. Homology	B. Multiple chromosome sets
3. Microevolution	C. Family tree
4. Extinction	D. Minor allele frequency changes
5. Hybrid sterility	E. Large phenotype changes
6. Macroevolution	F. Geographic separation
7. Species	G. Same ancestral origins
8. Allopatric speciation	H. A functional comparison
9. Polyploidization	I. No one breeds mules
10. Phylogeny	J. Total species death

Cell Biology True or False Quiz: Determine whether the following statements are true or false based on your understanding of cell biology concepts. Write "True" if the statement is correct and "False" if the statement is incorrect. Good luck!

1. ----- Temporal isolation functions as a postzygotic mechanism.

2. ----- Hybrid sterility represents a postzygotic mechanism.

3. ----- Mules demonstrate heterozygote advantage.

4. ----- Subspecies do not possess the genetic superiority of full species.

5. ----- The allopatric speciation model is rooted in geographic factors.

6. ----- Autopolyploids emerge as a result of hybridization.

7. ----- Carbon-14 dating lacks reliability beyond 40,000 years.

8. ----- A phylogeny delineates lineages of descent.

II. Completion Exercise: Complete each sentence by filling in the blanks with the appropriate term or phrase. Write your answers in the space provided. Good luck!

1. _____ arises when two remotely related lineages develop structures that exhibit superficial similarities.

2. _____ describes the scenario in which two closely related lineages evolve structures that display both superficial and genuine similarities.

3. The wing of a grasshopper and the wing of a bird, both serving for flight, are hence considered _____.

4. The wings of a bird and a bat, serving for flight and stemming from a common ancestor, are thus deemed _____.

5. The permanent disappearance of a species is termed _____.

6. The concept suggesting that evolution unfolds through rapid changes followed by extended periods of stability is referred to as _____.

III. Multiple Choice: Choose the best answer for each question by circling the corresponding letter.

1. The contemporary understanding of a species relies on_____.

a. Physical and biological traits

b. Shared lineage

c. Biochemical resemblances

d. Reproductive barriers

e. Morphological likenesses

2. All the given options represent prezygotic isolating mechanisms except for _____.

a. Ecological isolation

b. Hybrid sterility

c. Behavioral isolation

d. Mechanical isolation

e. Temporal isolation

3. Speciation in Hawaiian sap-feeding *Drosophila* occurred due to _____.

a. Ecological isolation

b. Behavioral isolation

c. Inbreeding isolation

d. Postzygotic isolation

e. Hybrid inviability

4. According to Mayr's theory of speciation, the initial step in the formation of a new species is _____.

a. The emergence of new genes

b. The onset of reproductive isolation

c. An increase in hybridization

d. The development of novel morphologies

e. The establishment of a barrier to gene flow

5. Two distinct plant species produce hybrid seeds that fail to germinate. This scenario exemplifies _____.

a. Speciation

b. Hybrid vigor

c. Hybrid inviability

d. Hybrid sterility

e. Mechanical isolation

6. The majority of crop plant species have arisen through _____.

a. Allopatric speciation

b. Sympatric speciation

c. Polyploidization

d. Subspeciation

e. Punctuation

7. Changes in the course of a river divide a population of beetles into two isolated groups, which subsequently evolve into two distinct species. This scenario is an example of _____.

a. Sympatric speciation

b. Allopatric speciation

c. Peripatric speciation

d. Punctuated equilibrium

e. Macroevolution

8. Based on variations in the developmental patterns of human and chimpanzee skulls, some biologists suggest that speciation might result from mutations in _____.

a. Structural genes

b. Introns

c. Structural proteins

d. Regulatory genes

e. All of the above

9. The alterations in chromosome structures that arise in a population of grasshoppers and lead to the formation of two new species are an example of _____.

a. Gene flow

b. Macroevolution

c. Sympatric speciation

d. Allopatric speciation

e. Ecological isolation

10. All the following are instances of macroevolution except _____.

a. The evolution of stomata in plants

b. The evolution of terrestrial vertebrates

c. The emergence of homeostasis

d. Speciation in a population of grasshoppers

e. The development of the jawbone in fish

11. Populations of the same species residing at opposite ends of a geographic barrier are more prone to speciation than those centrally located due to _____.

a. Differing environmental conditions

b. Sympatry

c. Mechanical isolation

d. Behavioral isolation

e. Changes in chromosome structure

12. The process of constructing lineages from the fossil record and biological characteristics is known as _____.

a. Paleontology

b. Phylogeny

c. Taxonomy

d. Radioisotope dating

e. None of the above

13. The independent evolution of spiny mammals in Africa and South America exemplifies _____.

a. Parallel evolution

b. Convergent evolution

c. Sympatric speciation

d. Divergent evolution

e. None of the above

14. All the following are instances of homology (in evolutionary terms) except _____.

a. The human's arm and the bat's wing

b. The bird's wing and the human's arm

c. The whale's flipper and the human's arm

d. The bird's wing and the fly's wing

e. The horse's leg and the human's leg

15. The theory of punctuated equilibrium was proposed by _____.

a. David Raup

b. Eldridge and Gould

c. David Jablonski

d. M.J.D. White

e. Ernst Mayr

16. The most common evolutionary pattern is _____.

a. Divergence

b. Convergence

c. Parallelism

d. Relay

e. Microevolution

17. Gradualism refers to _____.

a. Mass extinctions

b. The climatic changes that cause mass extinctions

c. The slow rate of change in species over time

d. Gaps in the fossil record

e. None of the above

18. One of the challenges in applying the modern definition of a species is that _____.

a. It does not account for common ancestry

b. It does not apply to asexual organisms

c. It does not apply to geographically separated populations

d. It has limited applicability when dealing with sympatric populations

e. It does not account for ecological isolation

19. Adaptive radiation arises from _____.

a. The reduction of barriers to gene flow

b. The emergence of new ecological opportunities for a species

c. The restriction of a species' ecological range

d. The presence of similar species in overlapping areas

e. Hybridization

20. Two distinct species of plants produce a sterile hybrid that undergoes chromosome doubling and successfully self-fertilizes. A new species has been formed through _____.

a. Allopatry

b. Behavioral isolation

c. Polyploidy

d. Prezygotic isolation

e. None of the above

READING UNDERSTANDING

I.   Recent Insights into Tooth Decay

Extensive research into the causes and prevention of tooth decay has yielded some unexpected results. One notable discovery is that cheddar cheese may actually inhibit the progression of tooth decay. When consumed immediately after sugary foods, cheddar cheese appears to have a decay-retarding effect on human teeth. The underlying mechanism by which cheese exerts this protective effect remains unclear. Researchers speculate that cheddar might interfere with the acids responsible for tooth decay or the bacteria that produce these acids. If these hypotheses are confirmed, cheddar cheese would be the first common food identified to possess such a beneficial property.

Another intriguing finding from the research is that the sugar content in heavily sweetened cereals has a similar impact on tooth decay, regardless of whether the sugar content is eight percent or nearly eight times that amount. This suggests that once a certain threshold of sugar is surpassed, the risk of decay does not proportionately increase, indicating a more complex interaction between sugar and oral health than previously understood.

Academic Extension

The potential cariostatic (decay-preventing) effect of cheddar cheese aligns with broader research into functional foods, which are those that provide health benefits beyond basic nutrition. The hypothesis that cheese might neutralize harmful acids or inhibit acid-producing bacteria could open new avenues for dietary recommendations and public health strategies aimed at reducing dental caries. Investigations into the specific components of cheese, such as casein, calcium, and phosphate, could shed light on their individual roles in enhancing oral health.

Moreover, the finding regarding sweetened cereals suggests a need to revisit our understanding of how dietary sugars contribute to dental caries. The traditional belief that higher sugar concentrations directly correlate with increased decay risk may be overly simplistic. This indicates the necessity for further research into how different types of sugars, their physical forms, and the presence of other dietary components influence dental health. Such research could potentially inform the development of more nuanced dietary guidelines and preventive measures against tooth decay.

In conclusion, these findings emphasize the importance of exploring not just the quantity of sugar in foods, but also the interplay between various dietary elements and their collective impact on oral health. Understanding these complex interactions will be crucial for developing effective strategies to combat tooth decay and promote overall dental well-being.

1. Based on the passage, how many test outcomes were surprising?

A. None

B. Two

C. Four

D. Ten

2. Based on the passage, what impact does cheddar cheese appear to have?

A. It disrupts the function of teeth

B. It enhances the sweetness of sugar

C. It reduces the rate of tooth decay

D. It aids in food digestion

3. From the passage, it can be inferred that the research on the connection between cheese and teeth __.

A. Has been debunked

B. Will face significant delays

C. Has been proven definitive

D. Will proceed further

4. A good body of literature demonstrated that sweetened cereals were __ .

A. Nutritionally significant

B. All unexpectedly high in sugar

C. More costly than cheese

D. Equally damaging to teeth

II. Reproductive Strategies and Environmental Adaptations of Alligators

A. *Mating and Nesting Behavior of Alligators*

Research has shown that after mating, male and female alligators separate. Approximately one month post-mating, the female begins constructing a nest, typically on dry land, which resembles a small haystack. She meticulously gathers plants and other vegetation, layering them until the nest meets her satisfaction, frequently inspecting her work. Once the nest is prepared, she digs a shallow depression at the top, deposits around 40 eggs, and carefully covers them. Sixty-five days later, the hatchlings break through the eggshells. In some instances, the mother may assist by removing the top layer of the nest and occasionally even transporting the young to water in her mouth.

B. *Temperature-Dependent Sex Determination*

Temperature plays a crucial role in determining the sex of the hatchlings during the initial two to three weeks of incubation. Research indicates that an average temperature of 86 degrees Fahrenheit or lower results in a clutch of females, while a temperature of 93 degrees Fahrenheit or higher yields all males. Intermediate temperatures produce a mixed-sex litter. This phenomenon of temperature-dependent sex determination (TSD) is not unique to alligators but is also observed in other reptiles, including turtles and some lizard species. TSD is an adaptive strategy that can influence population dynamics and has significant implications for the conservation of species in the context of climate change.

C. *Maternal Behavior and Defense Mechanisms*

Maternal behavior among alligators varies significantly. Some female alligators aggressively defend their nests against potential threats, exhibiting strong territorial behaviors. In contrast, others are less defensive. Regardless of their initial defensive stance, adult alligators of both sexes generally respond swiftly to the distress calls of their young. The baby alligator's distress call, often described as a "yurk" sound, triggers an instinctual protective response from adults. This behavior, however, has been exploited by hunters who mimic the distress call to attract and capture adult alligators. Understanding these behavioral patterns is crucial for the development of effective conservation strategies and the management of alligator populations.

D. *Implications for Conservation and Climate Change*

The interplay between temperature and sex determination highlights the vulnerability of alligator populations to climate change. As global temperatures rise, the skewed sex ratios could lead to long-term population imbalances. Conservation efforts must consider these factors to ensure the sustainability of alligator populations. Additionally, protecting nesting sites and mitigating human-induced threats, such as hunting and habitat destruction, are essential components of effective conservation strategies.

In conclusion, the nesting behavior, temperature-dependent sex determination, and maternal instincts of alligators provide valuable insights into their reproductive strategies and the challenges they face. Further research is needed to understand the full implications of environmental changes on these ancient reptiles and to develop adaptive management practices that support their continued survival.

1. What is the primary focus of the passage?

A. How alligators select their partners

B. Events following alligator mating

C. The process of determining alligator sex

D. Alligator dietary habits

2. As per the text, how long does it take for alligator eggs to hatch?

A. 21 days

B. 55 days

C. 4 weeks

D. 65 days

3. When does an alligator's sex become evident?

A. During the mating process.

B. Approximately one week before birth.

C. At the moment of birth.

D. Within the initial weeks of life.

4. The term "clutch" most likely refers to which category of alligators?

A. Juvenile alligators

B. Mother alligators

C. Adult alligators

D. Male alligators

5. Among the given temperature ranges, which is most likely to result in a varied sex distribution among alligator hatchlings?

A. 84°F

B. 86°F

C. 90°F

D. 93°F

Reading Materials: How Species Form

To validate Darwin's theory of evolution, it is insufficient to merely demonstrate that evolution has occurred—for instance, that the ancestors of dinosaurs were fish and our own ancestors were apelike primates. It is essential to observe evolution in action. Although the complete replacement of one organism by another typically happens too slowly to witness within a human lifespan, less drastic changes within individual species can be observed. Many of these changes, when studied by biologists, reflect the selection of organisms better adapted to their environment, aligning with Darwin's proposal. The crucial question is whether these within-species changes lead to differences between species, ultimately resulting in the formation of new species.

If Darwin were alive today, he would be astonished by the vast amount of evidence supporting his theory that natural selection is the primary mechanism driving evolution. He wouldn't need further convincing, as the case he presented in "On the Origin of Species" is compelling. The current evidence for evolution stands firmly on three pillars: first, that natural selection drives microevolutionary change; second, that microevolutionary change leads to macroevolutionary change; and third, that macroevolution has indeed taken place.

In the previous chapter, we examined the evidence for macroevolution and established that natural selection is the agent of microevolutionary change or change within populations. We saw that natural selection drives evolution by favoring genetic variations that enhance an organism's ability to survive and reproduce. Different environments require different responses from the organisms living in them, resulting in a continuous process of adaptation to these environments. This is the key takeaway from the previous chapter: evolution is not random but directed by environmental pressures. Just as a football coach tests various plays but retains only those that succeed in the team's strategy, a population of organisms retains only the changes that are advantageous. The population does not choose which changes to keep, just as the coach does not decide which plays will work. The pattern of success determines the outcome.

1. The Nature of Species

How do adaptive changes within populations lead to the creation of new species? This question greatly intrigued Darwin, as he viewed species as the fundamental units of evolution. To delve into this topic, it is helpful to first explore what "species" means and how the concept has evolved over time.

John Ray, an English clergyman and scientist, was among the first to define "species." Around 1700, he suggested that a species could be identified by the ability of its members to interbreed and produce offspring that also belonged to that species. Even if offspring from the same mating looked different, they were still considered part of the same species. For example, all dogs, despite their varied appearances, belong to the same species because they can interbreed and produce viable offspring. Similarly, diverse forms of cabbages, such as cauliflower, rutabaga, and Brussels sprouts, are all interfertile and thus members of the same species, having been shaped by artificial selection. However, dogs and cats, trout and goldfish, and monkeys and humans, are all different species because they cannot interbreed to produce offspring.

Historically, people have always recognized different species. The term "species" itself is derived from the Latin word for "kind." John Ray's work helped elevate the species concept to an important biological category that could be classified and studied. He, along with his contemporaries, believed species were immutable—a view that persisted until Darwin's work in 1859. Darwin saw the existence of distinct species as critical evidence for his theory of evolution, explaining that each species occupies a specific role or niche in nature.

With the advent of Mendelian genetics in the early 20th century, there was a push to more precisely define species. In the 1930s, American evolutionist Ernst Mayr proposed that species were "groups of actually or potentially interbreeding natural populations which are reproductively isolated from other such groups." In essence, hybrids between species are rare in nature, while members of the same species can freely interbreed. However, this definition proved overly simplistic, as many groups of organisms, such as certain trees, mammals, and fish, can form fertile hybrids, though they may not commonly do so in nature.

In other cases, hybrids between species of a particular genus are generally sterile, or local populations within the same species may be unable to interbreed. This is seen in many annual plants. Additionally, for some plants and animals, it is still unclear whether they can form hybrids.

In practice, scientists typically identify species based on their distinct features rather than reproductive barriers. Consequently, modern biologists define a species as a group of organisms that are distinct from other such groups and do not typically interbreed extensively with them in nature, although they might under artificial conditions.

2.　*The Divergence of Populations*

The transition from microevolutionary changes within a species to the divergence of that species into two distinct species involves dividing a group of similar individuals into two less similar groups. These new species typically do not interbreed, either due to their inability to do so or because their occurrences prevent such interactions. This separation into dissimilar groups is called divergence, and the mechanisms preventing successful mating are known as reproductive isolating mechanisms. Let us first explore divergence.

Every species consists of local populations, groups of individuals that live together in a specific area and are more likely to interact with each other than with members of other populations. The effective breeding population, which includes members of a local population that actually interbreed, is often very small, leading to limited genetic exchange between different local populations. Consequently, local populations tend to adapt uniquely to the demands of their specific environments, resulting in changes in their characteristics.

The rate at which local populations change is primarily determined by the strength of the selective forces imposed by their local environments. Strong selective forces lead to rapid changes within local populations. Because the environments of local populations can be highly diverse, natural selection can cause these populations to adapt in significantly different ways. Thus, the adaptive changes within a population, or microevolution, lead directly to divergence between local populations. If this divergence continues far enough, it results in the formation of separate species from what originally were subpopulations of the same species. Therefore, extensive microevolution can lead to macroevolution.

Divergence can create new species, groups of individuals that are significantly different from each other. These new species may become so specialized to their local environments that they cannot survive elsewhere, preventing them from interacting with other species. Many tree and fish species exhibit this pattern. In such cases, the formation of species does not necessarily involve reproductive isolating mechanisms; instead, their specialized adaptations keep them apart.

3. *Reproductive Isolating Mechanisms*

Once microevolutionary adaptation leads to divergence, how do the resulting species maintain their distinct identities? In some cases, geographical isolation prevents gene exchange between isolated species, keeping them distinct. In other cases, changes occur to maintain these differences even if the isolated species come back into contact.

A. *Ecological and Behavioral Isolation*

Isolated populations may begin to exploit different resources in various ways. Due to their different approaches to the environment, these populations will remain distinct even if they migrate back into contact. This type of divergence is common: local populations might occupy different habitats, live in different types of places, complete different stages of their life cycle at different times, or develop different feeding habits. They may also exhibit different behaviors, resulting in a tendency not to mate with each other or simply not to come into contact. Any of these changes can lead to reproductive isolation. The factors involved in this form of reproductive isolation, which prevent the formation of zygotes, are called prezygotic factors. A zygote is the initial cell formed after the fusion of egg and sperm in a eukaryotic sexual organism. The zygote divides by mitosis and eventually develops into an adult organism.

B. *Genetic and Physiological Incompatibility*

Sometimes, isolated populations can no longer interbreed due to changes in their chromosome organization or physiological makeup, rather than differences in environmental adaptation. Even if individuals from two isolated populations mate, successful reproduction may not occur. If the genes from the two populations do not function harmoniously during development—a complex process—the hybrid individuals will not be viable. The factors involved in this type of reproductive isolation, which prevent the proper functioning of zygotes, are called postzygotic factors.

By understanding these mechanisms, we can see how reproductive isolation helps maintain the distinct identities of diverging species, ensuring their continued evolution and adaptation to their environments.

CHAPTER 9

Population Dynamics and Ecology

Abstract: Population dynamics is shaped by multiple factors influencing growth, distribution, and constraints. Key characteristics of populations include natality, mortality, and density. Exponential growth occurs under ideal conditions but is typically limited by carrying capacity, resulting in a logistic growth curve. Reproductive time lag affects population fluctuations, and age structure and reproductive strategies further influence growth rates. Populations are constrained by density-dependent factors such as predation and disease, and density-independent factors like natural disasters. Distribution patterns, including clumped, uniform, and random, are affected by interspecific interactions and competition. Allelopathy, resource partitioning, and character displacement illustrate competition in both terrestrial and aquatic environments. The exponential rise in human population highlights the urgency of sustainable resource management, with current growth rates potentially surpassing Earth's estimated carrying capacity.

Keywords: Allelopathy, Carrying capacity, Population dynamics, Reproductive strategies, Sustainable resource management.

UNDERSTANDING THE CHAPTER'S FOCUS: A ROADMAP

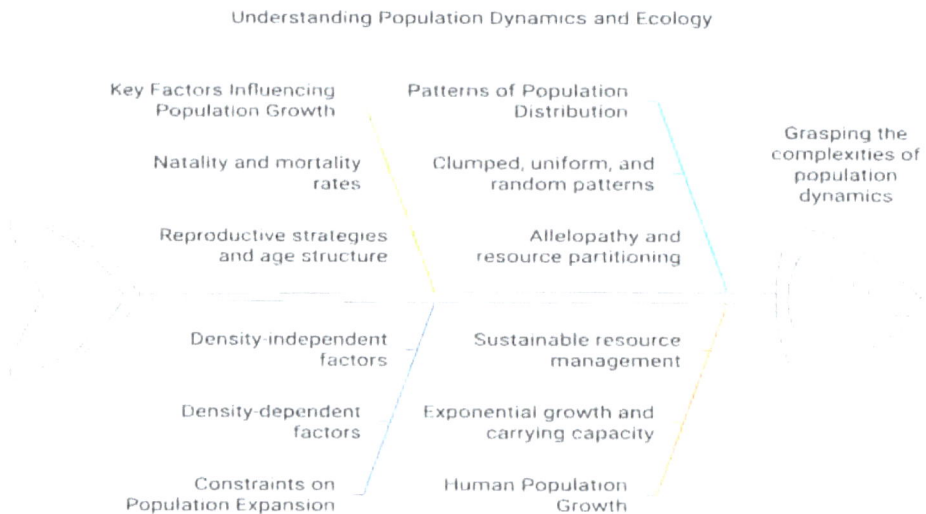

Understanding Population Dynamics and Ecology

Key Factors Influencing Population Growth

Natality and mortality rates

Reproductive strategies and age structure

Density-independent factors

Density-dependent factors

Constraints on Population Expansion

Patterns of Population Distribution

Clumped, uniform, and random patterns

Allelopathy and resource partitioning

Sustainable resource management

Exponential growth and carrying capacity

Human Population Growth

Grasping the complexities of population dynamics

Mohammad Mehdi Ommati

1. FACTORS INFLUENCING POPULATION GROWTH

Populations, defined as groups of individuals belonging to the same species, exhibit three key statistical characteristics: per capita birth rate (**natality**), per capita death rate (**mortality**), and population density (number of individuals per unit area).

As first described by Malthus, populations have the potential to grow exponentially if resources such as food and shelter are unlimited and if there are no threats from predation or competition. This type of growth is represented by an **exponential growth curve**. However, such ideal conditions rarely occur in nature. The finite availability of resources sets an upper limit to population size, known as the **carrying capacity** (K), which can be approached but not sustainably exceeded. **A logistic growth curve** depicts how population growth levels off as it reaches equilibrium with available resources.

When a population approaches or exceeds resource limits, there is a lag time before birth rates decrease and death rates increase. This response time is known as reproductive time lag, which contributes to the fluctuations in population numbers observed in natural populations. Seasonal changes often cause the carrying capacity and, consequently, population size to fluctuate. If a population significantly exceeds its environment's carrying capacity, it may cause lasting damage that permanently reduces the carrying capacity.

In addition to environmental carrying capacity, a population's growth rate is influenced by its age structure and reproductive strategy. **Age structure** reflects the proportion of young, middle-aged, and older individuals within a population. Populations with many individuals at or near reproductive age are likely to grow significantly. The age structure can also be depicted by a **survivorship curve**.

Reproductive strategies of populations are complex and have evolved over millennia. Generally, these strategies fall into two categories: R-selected and K-selected species. R-selected species mature quickly and produce many offspring, each small and equipped with few resources. Only a few offspring survive to reproductive age. K-selected species mature slowly and produce few offspring, but parents invest significant resources in each one, enhancing their survival chances until reproductive age.

2. CONSTRAINTS ON POPULATION EXPANSION

Population size is typically measured by its density. Regardless of whether **population density** is high or low, individuals within the population are usually unevenly distributed. Common distribution patterns include clumped, uniform, and random. High population density often leads to negative consequences and **density-dependent factors**, such as increased predation, parasitism, disease, and competition (both **intraspecific** and **interspecific**). Population size can also be affected by **density-independent factors**, such as natural disasters.

Predator-prey interactions affect population size in complex ways. These populations may cycle regularly between growth and decline, partly due to reproductive time lags. Predation can slow or halt prey population growth only when many reproductive individuals are removed. If predators mainly target the weak, sick, or young, the overall effect on population density may be minimal.

Ecologists debate whether species diversity in a community promotes stability or if stability fosters species diversity. One aspect of this debate is the hypothesis that complex food webs are more stable than simple ones. However, many stable, diverse communities in nature feature numerous simple food webs. It might be that stable environments foster diversity by allowing rare species to persist.

3. PATTERNS OF POPULATION DISTRIBUTION

Competition, predation, and other factors interact to determine a population's size within a community and its distribution. The distribution of a population within its potential range depends on the availability of food, suitable habitat, interspecific competition for resources, and other variables. In plants, one effective form of interspecific competition is **allelopathy**, where chemicals are released to inhibit competitors (Fig. **1**). Resource partitioning is common among species sharing similar habitats. In **character displacement**, closely related species evolve physical differences to exploit limited resources differently. Over time, these adaptations may lead to speciation. Similar to terrestrial plants, many aquatic plants, and algae release substances into the water that can affect other aquatic organisms. These chemicals can suppress the growth of competing plants or algae and sometimes affect other aquatic animals. However, it is not exclusive to plants, aquatic plants, or algae. Allelopathy can also occur in some animals, although it is less common and often referred to by different terms such as chemical defense or chemical communication.

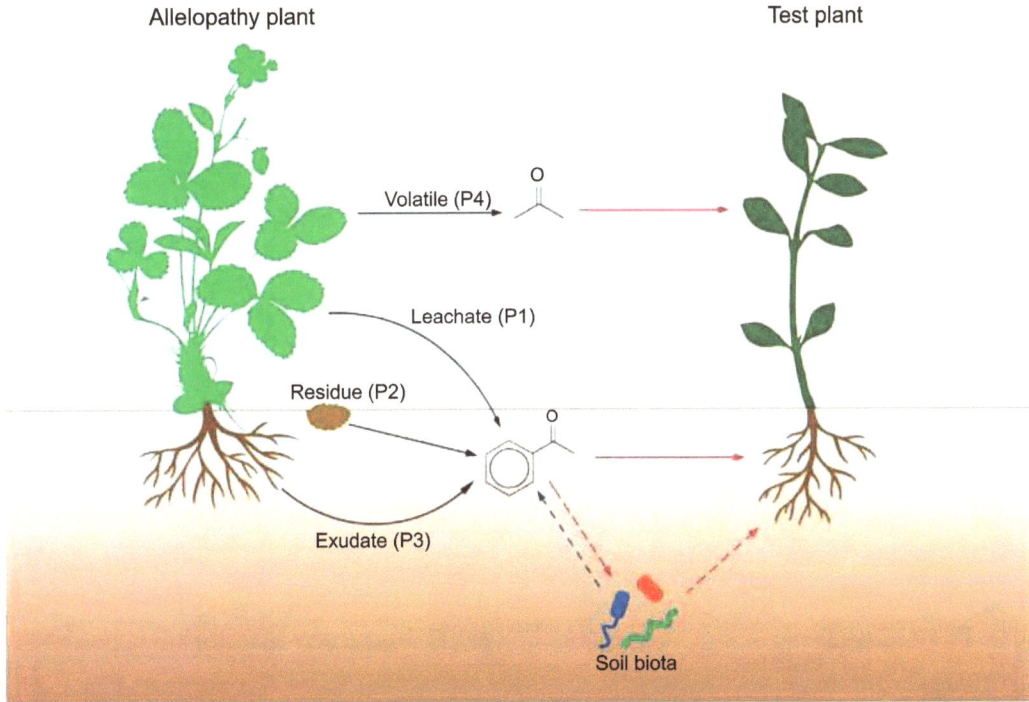

Fig. (1). Effect of Allelopathy on Plant Performance. Allelopathy is primarily a phenomenon observed in plants, where certain plants release chemicals into the environment that can inhibit the growth of other nearby plants. However, the concept of allelopathy has also been studied in aquatic ecosystems, including freshwater environments. In freshwater ecosystems, allelopathy can occur between aquatic plants, algae, and microorganisms.

In animals, similar chemical interactions are often categorized differently:

- **Chemical Defense:** Some animals release chemicals to deter predators or competitors. For example, skunks spray a noxious liquid to fend off threats.

- **Chemical Communication:** Many animals use pheromones to communicate with others of their species, influencing behaviors like mating or marking territory.

Examples of Animal Allelopathy-like Phenomena

a. Marine Animals: Some marine organisms, such as certain species of sponges, corals, and sea slugs, produce chemicals that inhibit the settlement and growth of competing species.

b. Insects: Some ants release formic acid as a defense mechanism and inhibit plant growth around their nests to reduce cover for predators (Fig. **2**).

Fig. (2). Woodland ants in Dorset fire jets of formic acid into the air to deter aerial predators when threatened.

While these examples might not strictly fall under the traditional definition of allelopathy, they illustrate that chemical interactions affecting other organisms' growth or behavior are present across a variety of life forms, including animals.

4. HUMAN POPULATION GROWTH: AN EXPONENTIAL CASE STUDY

The remarkable rate of increase in the human population began around 10,000 years ago with the agricultural revolution. At that time, the world's population was approximately 133 million; today, it has surged to over 5 billion. This increase follows an exponential growth curve, a pattern that is unsustainable.

While birth rates in many developed countries have slowed, they remain high in less developed regions. If this trend continues, demographers predict the human population could reach 30 billion by the end of the twenty-first century. However, most biologists estimate the Earth's carrying capacity for humans to be around 10 billion. This discrepancy highlights the urgent need for sustainable resource management to support the growing human population.

Vocabulary (Index of Terms)

Age Structure [eid3 'strʌktʃa]	In a population, the distribution of individuals across different age categories.
Allelopathy [ə‚liːˈlɒpəθi]	A chemical interaction between organisms where one organism inhibits the germination, growth, or reproduction of another by releasing toxins into the environment. This phenomenon occurs particularly among aquatic ecosystems, including freshwater environments; flowering plants, bacteria, and fungi.
Carrying capacity [ˈkæriŋ kəˈpæsɪti]	The maximum number of individuals in a population (or species) that can be sustained indefinitely by a given environment.
Character displacement [ˈkærɪktər dɪsˈpleɪsmənt]	An alteration of the traits of a species as a result of competition or other interactions with associated species.
Density-dependent factors [‚dɛnsɪti-dɪˈpɛndənt ˈfæktərz]	Population-limiting factors that become more effective as the size of the population increases.
Density-independent factors [‚dɛnsɪti-ɪnˈdɛpəndənt ˈfæktərz]	Population-controlling factors that are not related to the size of the population.
Exponential growth curve [‚ɛkspəˈnɛnʃəl ˈgroʊθ kɜːrv]	A pattern of population growth in which greater and greater numbers of individuals are produced during successive doubling times; the pattern that emerges when the per capita birth rate remains even slightly above the

per capita death rate, putting aside the effects of immigration and emigration.

Interspecific competition
[ˌɪntərspəˈsɪfɪk ˌkɑːmpəˈtɪʃən]

Two-species interaction in which both species can be harmed due to overlapping niches.

Intraspecific competition
[ˌɪntrəspəˈsɪfɪk ˌkɑːmpəˈtɪʃən]

Interaction among individuals of the same species that are competing for the same resources.

Logistic growth curve [ləˈdʒɪstɪk ˈɡroʊθ kɜːrv]

A pattern of population growth in which a low-density population slowly increases in size, goes through a rapid growth phase, and then levels off once the carrying capacity is reached.

Mortality [mɔːrˈtælɪti]

The number of individuals leaving the population by death per thousand individuals in the population.

Natality [neɪˈtælɪti]

The number of individuals entering the population by reproduction per thousand individuals in the population.

Population density [ˌpɒpjʊˈleɪʃən ˈdɛnsɪti]

The number of individuals of a population that are living in a specified area or volume.

Survivorship curve [səˈvaɪvərʃɪp kɜːrv]

A plot of the age-specific survival of a group of individuals in a given environment, from the time of their birth until the last one dies.

COMPREHENSION TASKS

I. Key Terms: Matching

Instructions: Match each term from the left column with the corresponding description from the right column. Write the letter of the correct description next to the number of the term. Each description will only match with one term. Good luck!

1.	Natality	**A.**	k
2.	K-selected species	**B.**	Per capita birth rate
3.	Interspecific competition	**C.**	Reproduce many young
4.	Allelopathy	**D.**	Between species
5.	Population density	**E.**	Reproduction tied to k
6.	Exponential growth	**F.**	Within species
7.	R-selected species	**G.**	Chemical warfare
8.	Mortality	**H.**	Individuals per area
9.	Carrying capacity	**I.**	"no limits"
10.	Intraspecific competition	**J.**	Death rate

Cell Biology True or False Quiz: Determine whether the following statements are true or false based on your understanding of cell biology concepts. Write "True" if the statement is correct and "False" if the statement is incorrect. Good luck!

1. ----- The maximum sustainable population in an ecosystem cannot be surpassed indefinitely.

2. ----- Logistic growth results in a sigmoidal curve shape.

3. ----- Species with R-selected traits typically lay eggs with substantial yolk reserves.

4. ----- The distribution of a horse herd appears random.

5. ----- Examples of density-independent factors include competition and predation.

6. ----- Allelopathy encompasses the synthesis of harmful substances.

7. ----- Galapagos finches display distinct morphological differences in response to ecological competition.

8. ----- Exponential growth is not sustainable over extended periods in natural populations.

II. Completion Exercise: Complete each sentence by filling in the blanks with the appropriate term or phrase. Write your answers in the space provided. Good luck!

1. The term used to describe the quantity of individuals within a given space is the _____ of a population.

2. A visual representation illustrating the distribution of survivors across various age categories within a population is termed a _____.

3. _____ competition exclusively concerns individuals within the same species.

4. The duration needed for birth rates to decline and death rates to increase is referred to as the _____.

5. Species that are _____ selected develop gradually and yield a smaller number of offspring.

6. Animals that establish territories exhibit a _____ dispersal arrangement.

III. Multiple Choice: Choose the best answer for each question by circling the corresponding letter.

1. Within the exponential growth equation, the symbol dN / dt represents __.

a. Population size

b. Rate

c. Time

d. Alteration in population size

e. None of the above

2. Parasites and pathogens spread most swiftly in a population when: _____.

a. The host population is randomly distributed

b. The host is well-adapted to them

c. The parasite is highly virulent

d. The host population is densely packed

e. The host is debilitated

3. In the absence of limiting resources, populations expand _____.

a. Exponentially

b. Logistically

c. At a diminished pace

d. At a fluctuating pace

e. At none of the above rates

4. The final segment of the logistic growth curve signifies _____.

a. The phase of rapid expansion

b. The stage where the population has reached its maximum capacity

c. The point at which K - N = 0

d. The segment beneath the carrying capacity

e. Options b and c

5. All of the subsequent factors can influence the carrying capacity of a population except _____.

a. Seasonal fluctuations in resources

b. A decrease in nesting sites due to logging activities

c. A shortage of food due to floods

d. A shift in the birth rate

e. Alterations in habitat

6.   In a rapidly expanding population: _____.

a. The number of youthful individuals surpasses the number of elderly individuals

b. The number of older individuals surpasses the number of young

c. There is an equal distribution of age

d. There is a constant distribution of age

e. The birth rate matches the death rate

7.   Every one of these traits is typical of R-selected species _____.

a. Minimal investment in offspring

b. Production of vast quantities of gametes

c. Elevated mortality rate among juveniles

d. Commencement of reproduction at a comparatively advanced age

e. Competitively disadvantaged juvenile stages

8.   Competing among distinct species for a finite resource is known as _____.

a. Interspecific competition

b. Scramble competition

c. Diffuse competition

d. Intraspecific competition

e. Density-independent competition

9. The description of the population of dandelions thriving in your garden applies to all statements except: _____.

a. They exhibit a random distribution

b. They possess effective dispersal capabilities

c. They generate large quantities of seeds

d. They are a K-selected species

e. Their resources are typically evenly dispersed throughout their habitat

10. Closely related species that exploit the same restricted resources frequently develop distinct characteristics through a process termed _____.

a. Allopatric speciation

b. Sympatric speciation

c. Character displacement

d. Allelopathy

e. Resource partitioning

11. Density-independent factors may encompass _____.

a. Fluctuations in seasons

b. Unforeseen accidents

c. Adverse weather conditions

d. Natural calamities

e. All of the aforementioned

12. In Gause's experiments on predator-prey dynamics, providing a hiding place for prey resulted in: _____.

a. Extinction of the predator

b. Extinction of the prey

c. Cycling of both predator and prey populations

d. Cycling of only the prey population

e. Cycling of the prey population and extinction of the predator

13. The emission of a chemical by a plant species that is harmful to another plant species is termed _____.

a. Character displacement

b. Intraspecific competition

c. Allelopathy

d. Resource partitioning

e. Homeopathy

14. Three species of warblers inhabit the same environment but forage at different heights in the canopy. This exemplifies _____.

a. Character displacement

b. Resource partitioning

c. Intraspecific competition

d. Density-independent competition

e. Allelopathy

15. The characteristic form of the logistic growth curve is _____.

a. J-shaped

b. S-shaped

c. A linear graph

d. A vertical line

e. L-shaped

16. If the human population continues to increase at its current pace, the projected population size by the end of the twenty-first century is estimated to be _____.

a. 10 billion

b. 200 million

c. 30 billion

d. 133 million

e. 6 billion

17. A population typically exhibits a clumped distribution pattern if: _____.

a. Individuals are territorial

b. Resources are abundant and patchy

c. Resources are thinly dispersed

d. Resources are evenly distributed

e. Resources are scarce

18. Density-dependent factors comprise intraspecific competition and _____.

a. Predation

b. Parasitism

c. Intraspecific competition

d. Disease

e. All of the above

19. A disturbance significantly reduces the population density of a species in a community. This reduction is expected to have the most substantial consequences if: _____.

a. The community is diverse

b. The community is simple and consists of very few species

c. The community contains 500 species

d. It is a tropical community

e. It is a marine community

20. All of the following are examples of K-selected species except _____.

a. An oak tree

b. An elephant

c. A red snapper

d. A Douglas fir tree

e. A giant panda

READING UNDERSTANDING

I. Lichens: Nature's Hidden Wonders

Lichens possess a remarkable beauty for those who take the time to observe, yet their understated appearance often leads them to be overlooked. While they may seem unassuming with their modest hues and apparent flatness as they adhere to various surfaces, they thrive in the background, seemingly crafted to evade notice. Nonetheless, they captivate botanists due to the unresolved mysteries they present and their multifaceted capabilities.

To the casual observer, the true nature of lichens remains concealed; few would suspect them to be a complex amalgamation of interacting life forms. Contrary to their outward simplicity, lichens comprise a symbiotic relationship between a fungus and a colony of algae (or blue-green algae, which some scientists classify as bacteria). Certain species even incorporate all three of these diverse life forms.

Lichens stand apart from their individual components in both appearance and biochemistry; many produce distinct compounds inaccessible to the constituent organisms alone.

Thriving in nearly every imaginable natural habitat, from arid deserts to lush tropical rainforests, lichens exhibit remarkable adaptability. They can be found in unexpected places, such as on the backs of specific beetles in New Guinea or within rocks alongside algae in the otherwise desolate dry valleys of Antarctica. While many species endure extreme conditions of heat, cold, or aridity, only a few withstand heavy air pollution, often thriving in environments where the air is exceptionally pure. The absence of lichens from an area serves as an ominous indicator of environmental peril.

1. What would be the most fitting title for the passage?

A. Exploring the Versatility and Complexity of Lichens

B. Unveiling the Enigmatic Traits of Algae Colonies

C. Investigating the Decline of Lichen Species

D. Discovering the Diverse Habitats of Spectacular Fungi

2. The author suggests that lichens grow "as though designed to be ignored" because they are _____.

A. Not fully comprehended by botanists

B. Challenging to collect for study purposes

C. Uncomplicated in their internal structure

D. Not readily noticeable by observers

3. According to the author, most individuals are unaware that lichens are _____.

A. Leafy plants

B. A class of simple bacteria

C. Two-dimensional life forms

D. Combinations of organisms

4. All of the following are mentioned as components of lichens EXCEPT _____.

A. A fungus

B. Algae

C. A beetle

D. Bacteria

5. The "unique" compounds referenced in the second paragraph are produced ___.

A. Through the collaborative efforts of the lichen's constituents

B. Solely under laboratory conditions

C. Through one of the three potential processes

D. Intermittently throughout the lichen's life cycle

6. The primary focus of the third paragraph is _____.

A. Describing the appearance of lichens

B. Delineating the habitats where lichens thrive

C. Explaining the classification of lichens

D. Elucidating the reproduction of lichens

7. The author suggests that lichens could potentially be employed to _____.

A. Locate water sources

B. Eradicate unwanted vegetation

C. Assess air quality

D. Provide sustenance in remote regions

II. Exploring the Diverse Inhabitants of the Bayous

In the waters of the bayous, a myriad of creatures beyond fish thrive. Clams and snails make their homes by burrowing into the bottom sediment or clinging to aquatic plants. Crabs and shrimp migrate upstream from the bays, while their freshwater counterparts, the crayfish, inhabit nearly every water body, from ponds to puddles.

Among the bayou dwellers are mammals like beavers, which lead predominantly aquatic lives, often residing within the banks of the waterways rather than constructing conspicuous dams or lodges. These elusive creatures emerge at night to forage, leaving behind telltale signs such as felled trees and stripped bark.

Playful otters navigate the bayous with ease, equally adept in water and on land. Feasting on fish, frogs, and crayfish, they share the aquatic domain with their relatives, the minks, which also find sustenance in the bayou waters.

In addition to mammals, muskrats erect their dome-shaped dwellings throughout the marshes, now sharing territory with the larger nutria, introduced to the Gulf Coast from South America. These sizable rodents, known for their appetite for vegetation, earn disfavor by venturing into agricultural fields, particularly rice and sugarcane crops.

1. What is the primary aim of the passage?

A. To provide a comparison of various rodent species

B. To delineate the inhabitants of the bayous

C. To explain the construction of aquatic animal habitats

D. To examine the flora of the southern United States

2. According to the text, how do shrimp and crabs reach the bayous?

A. By leaping between ponds and puddles

B. By burrowing along the bank bottoms

C. By swimming upstream from the bays

D. By adhering to drifting plants' undersides

3. In the first paragraph, the term "cousins" refers to which of the following?

A. Bays

B. Crabs

C. Shrimp

D. Crayfish

4. To locate beavers in the South, one should look for which of the following?

A. Trees devoid of bark

B. Large dams spanning bayous

C. Stick dwellings encrusted with clinging snails and clams

D. Uncommon water plants thriving in bayous

5. From the passage, it can be inferred that beavers and otters are well-suited to bayou life because they _____.

A. Consume vegetation

B. Are aquatic animals

C. Possess dense fur

D. Construct substantial dams

6. The otter is closely related to which of the following?

A. The mink

B. The beaver

C. The muskrat

D. The crayfish

7. The author suggests that people have an aversion to nutrias because they
_____.

A. Resemble oversized rats

B. Construct dome-shaped habitats

C. Consume sugarcane and rice crops

D. Displace the more favored muskrat population

READING MATERIALS

Insights into Behavioral Ecology

Behavior is a fundamental aspect of an organism's life, influencing its survival, reproduction, and interaction with the environment. While many behaviors are inherited and evolved over time, others are learned through experience, contributing to the adaptability and success of a species. Understanding the mechanisms behind behavior sheds light on its ecological and evolutionary significance.

1. Understanding Behavior

Behavior encompasses the actions and reactions of organisms, playing a crucial role in their ecological roles and interactions. Unlike physical structures, behaviors are transient and challenging to study in the fossil record. Nevertheless, they are essential for tasks such as predator avoidance, mate selection, and environmental adaptation.

2. Instinct

Instinctive behavior, genetically inherited and preprogrammed, is vital for survival and reproduction. It allows organisms to respond appropriately to stimuli without prior experience, ensuring correct and necessary actions. While instinctive behaviors lack adaptability, they are effective in fundamental activities, particularly in species with short life cycles and limited parental care.

3. Learned Behavior

Learning, a change in behavior due to experience, complements instinct in many organisms, particularly those with longer life spans and parental care. Birds, for example, learn parts of their songs, blending innate abilities with acquired skills. Learning enhances adaptability, enabling animals to adjust behaviors to local conditions and environmental changes.

4. *Conditioning*

Classical conditioning, exemplified by Pavlov's experiments with dogs, demonstrates how organisms associate neutral stimuli with natural stimuli, leading to learned responses. Such conditioning aids in adapting to the environment, helping animals avoid harmful situations or locate resources.

5. *Imprinting*

Imprinting, observed in young animals, involves rapid learning of specific behaviors during critical periods. Ducklings, for instance, imprint on the first moving, noisy object they encounter, typically their mother, facilitating essential learning and survival skills.

6. *Insight Learning*

Insight learning involves problem-solving based on past experiences, and reorganizing knowledge to address novel challenges. While difficult to study, examples like Japanese macaques washing food suggest animals' ability to devise innovative solutions through insight.

Behavioral ecology draws from various disciplines to explore the ecological and evolutionary significance of behavior. By understanding how organisms generate and adapt behaviors, researchers can uncover the intricate ways in which animals navigate their ecological niches and ensure species survival and success.

Section 4
Advances in Biological Research and Information Systems

Biology for Students, 2025, 221-251

CHAPTER 10

Navigating Cancer Systems Biology

Abstract: Cancer systems biology integrates experimental models, data analysis, and dynamic network modeling to elucidate the complex mechanisms underlying cancer progression. This chapter outlines the essential requirements for experimental models, emphasizing the need for well-characterized cancer subtypes and high-quality mouse models that mimic clinical outcomes. It discusses various approaches to constructing cancer gene networks, including inference from genome-wide datasets, extension of protein interaction networks, and integration of high-throughput data with literature. The chapter also highlights advancements in bioinformatics, such as pattern recognition and machine learning, and the evolution of network visualization from static to dynamic models. Finally, it examines network analysis techniques for understanding biological systems and applying dynamic network modeling to decipher information processing in cancer cells. Data quality and model development challenges are noted, with a call for enhanced training in network-based thinking to further cancer research.

Keywords: Bioinformatics, Cancer networks, Data integration, Dynamic modeling, Network visualization.

UNDERSTANDING THE CHAPTER'S FOCUS: A ROADMAP

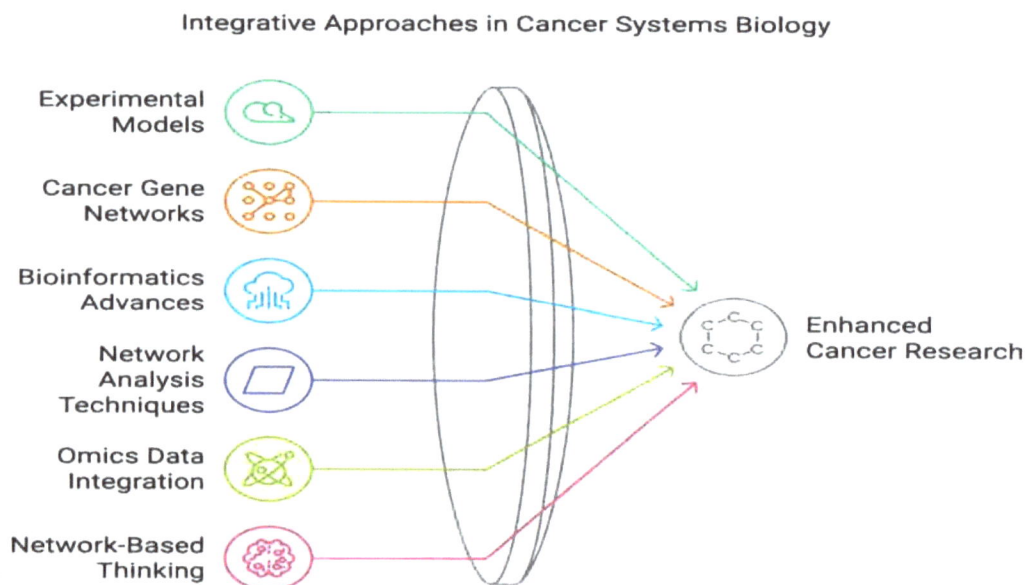

Integrative Approaches in Cancer Systems Biology

Mohammad Mehdi Ommati

1. CANCER SYSTEMS BIOLOGY AND ITS IMPACT ON PRECISION MEDICINE

1.1. The Transformative Impact of Systems Biology on Cancer Research

When a collision disrupts a congested road during rush hour in a bustling city like Tehran, Beijing, Toronto, or New York, traffic comes to a temporary standstill. Soon, however, drivers find alternative routes to reach their destinations. Similarly, the city's road map—a web of interconnected routes—allows for alternative paths. Increasing evidence indicates that cellular molecules are networked in a similar manner. This interconnected structure enables cancer cells to evade the effects of drugs.

Traditional biological investigations usually use straightforward reasoning and descriptions without any math. These methods work well only for simple processes that involve just a few parts or short cause-and-effect links. As a result, traditional biology finds it challenging to understand the complicated interactions between different molecules that are typical of many diseases, such as cancer. Cancer involves numerous interacting elements forming complex networks with highly nonlinear dynamics. Thus, focusing on just one molecule in a biochemical pathway often doesn't work well for treating cancer, since cells can discover other ways to work around the blockage. This challenge emphasizes why numerous existing drug designs do not achieve success and underscores the importance of adopting a systems perspective rather than a gene-centric approach in developing new cancer treatments.

Systems biology integrates empirical data with mathematical and computational techniques to understand complex biological occurrences. For instance, a multitude of proteins can play a role in signaling processes that are crucial for cell function. Disruptions in these networks can lead to cancer. Systems biology provides detailed maps of these cellular networks and utilizes sophisticated quantitative, statistical, and computational techniques to examine them. By understanding these complex systems, more effective therapeutic strategies can be developed, such as targeting multiple key points in a biochemical network simultaneously. This approach promises significant advancements in cancer treatment, transforming traditional reductionist methods into comprehensive systems-level strategies for drug discovery.

The emergence of systems biology is fueled by advancements in large-scale techniques designed for life science applications. Innovative techniques like next-generation genome sequencing, RNA-seq, ChIP-seq, and microarrays enable researchers to assess gene expression and regulatory elements across the entire genome. These advancements have transformed biological research from a "single gene model," which emphasizes individual genes and proteins, to a "multiple gene model," acknowledging that biological entities function as systems composed of interconnected components. With the decreasing cost and greater accessibility of these technologies, large-scale biological research projects are becoming more prevalent.

The rise of systems biology has significantly increased the amount of biological data, and this growth is expected to persist in the future. Large-scale data generated through advanced techniques is both extensive and impartial, greatly transforming cancer research. Scientists now face the challenge of managing, interpreting, and extracting insights from vast datasets. In this huge amount of data, important signals and biological patterns can be hidden, so using math and computers is key to finding them.

Systems biology tackles these problems by combining different types of omics data and creating computer tools to understand complex systems. It uses network and graph theory to mathematically explain, study, and model these biological systems. By applying network theory, biological language is transformed into a computable mathematical language, capable of handling the vast number of relationships in biological data. The foundation of systems biology is network biology, which uses networks to represent, analyze, and model biological complexity, aiming to reveal key biological principles.

This chapter presents various strategies, methodologies, and computational approaches used to study cancer systems biology, with an emphasis on network reconstruction, analysis, and modeling of biological networks. Relevant chapters of this book will guide readers through these strategies and procedures. Additionally, the chapter discusses the obstacles and difficulties faced in cancer systems biology.

1.2. Systems Biology as a Tool for Customized Medicine

New investigations reported that many medications work for less than half of the prescribed patients. In addition, around 3 million prescriptions are either incorrect or ineffective each year, leading to over 100,000 deaths in the U.S. from drug-related complications. These numbers highlight the shortcomings of a one-size-fits-all approach to medicine and preventive care. It is essential to consider the individual's unique genetic background to treat diseases effectively. Precision medicine aims to develop treatment plans customized to each individual's genetic profile, targeting the specific needs of various genetic subgroups. This method enables doctors to more accurately assess the risk-benefit ratio and prescribe suitable medications for different patient groups.

In the last ten years, cancer treatment has transitioned from a universal approach to a more individualized strategy. In individualized medicine, treatment is customized to each patient's unique genetic mutations found in their tumors. Nevertheless, cancer's complexity and heterogeneity present significant challenges. Advances in cancer treatment, within the context of individualized medicine, are expected to depend significantly on the cutting-edge scientific methods offered by systems biology. Consequently, extensive work in omics and systems biology has been carried out by the cancer research community, leading to significant funding directed toward cancer studies in recent years. This has led to the production of more large-scale data in cancer biology than in any other area of biological research. However, cancer's complexity still poses challenges to fully grasping the molecular mechanisms behind tumorigenesis. To tackle this issue, network-based methods have been created and implemented to study cancer's cellular networks.

Analyzing the entire tumor genome to pinpoint cancer driver mutations, along with comprehensive profiling of omics data—such as gene expression, epigenetic changes, metabolic information, and signaling data like phosphoproteomic profiles—will help build personalized cancer signaling networks for each patient. By examining these tumor signaling networks, more accurate, personalized risk assessments and treatment plans can be developed. The cost of sequencing a human genome is also dropping significantly, with emerging DNA sequencing technologies targeting a price point of $1,000 per genome. The lower costs now allow for the identification of key mutation genes specific to each patient, and profiling tumor gene expression is becoming more commonplace and readily available.

With these data routinely generated in clinical settings, adopting a systems biology approach for medical research is now achievable, bringing us closer to personalized medicine. For instance, creating and examining tumor signaling maps customized for individual patients can reveal crucial protein communication networks that are vital for the growth of particular tumors. Modeling and simulating these individualized tumor signaling maps aid in understanding the molecular processes involved in cancer, identifying critical tumor targets, and minimizing the likelihood of treatment failure during clinical phases. As a result, network analysis and modeling are anticipated to become essential tools in the pharmaceutical and biotech sectors. Three key areas of cancer biology are likely to gain substantial advantages from the integration of systems biology:

1. **Identification of Prognostic and Drug-Response Biomarkers:** Utilizing a systems-based approach to connect genomic information with clinical data (including blood tests, lifestyle surveys, and patient outcomes) enables the discovery of vital biomarkers.

2. **Exploring Network-Based Molecular Mechanisms:** Developing networks and computational models for various stages of cancer progression will deepen our insight into the molecular processes driving cancer.

3. **Understanding the Network-Driven Molecular Mechanisms of Metastasis:** Comparing the networks between primary and metastatic tumors can enhance treatment strategies for advanced-stage cancers and shed light on why some drugs are ineffective.

In conclusion, cancer systems biology provides fresh insights into network-driven molecular mechanisms, helping to select combinations of anti-cancer drugs and fine-tune treatment strategies for better outcomes.

2. APPROACHES TO INVESTIGATING CANCER SYSTEMS BIOLOGY

Recent tumor genome sequencing projects have uncovered thousands of cancer-driving mutations. These genes exhibit vast variability, with little commonality across different tumors. This diversity appears both across various types of tumors and within tumors that arise from the same tissue. The mentioned findings propose that cancer can arise from multiple genetic pathways. However, several key functional modules, called the hallmarks of cancer, have been recognized and well-documented. Genes that cause cancer mutations are often linked to these hallmarks or functional groups. Each hallmark includes a group of related pathways, making it possible to map these modules and mutation-causing genes onto smaller networks of interconnected pathways.

By combining analyses of the human signaling network with cancer driver-mutating genes, researchers have identified these network modules. The systems perspective in cancer research focuses on developing accurate network models of tumors while identifying network modules, crucial genes, and other significant characteristics within each module. Ultimately, the findings from this systems biology perspective need to be confirmed through experiments involving cancer cell lines (*in vitro*) and laboratory animal (*in vivo*) models. By combining biological and clinical data across different levels, cancer systems biology offers a profound understanding of this complex disease.

Network construction focuses on creating functional networks that show how genes and proteins interact under certain conditions, like in cancer signaling pathways related to metastasis. These networks capture the connections between different types of biological data and key processes involved in cancer growth and spread, such as the hallmarks of cancer, how the cell cycle is controlled, apoptosis (programmed cell death), and immune responses. By building networks that include data collected over time, researchers can better understand the dynamic changes that occur during tumor progression. This approach simplifies complex biological interactions, making it easier to analyze and identify how inputs and outputs relate to each other. Additionally, creating these networks involves using methods that combine various types of biological data with established scientific knowledge.

Network visualization offers tools that enhance intuition beyond what analytical tools alone can offer, aiding in forming hypotheses for further network exploration. Network analysis involves using mathematical and statistical methods to study constructed networks. This process helps identify key components, such as important nodes and interconnected groups, along with more complex interactions like teamwork, simultaneous expression, or shared regulation between modules. By analyzing these networks, researchers can uncover insights into how cancer operates, leading to hypotheses about the mechanisms that drive cancer progression and metastasis. Examining time-course networks helps us understand how cellular functions change (or fail) as diseases progress. This analysis shows how different components interact over varying time frames, highlighting the complexity and unpredictability of biological systems. Furthermore, network analysis can reveal gene signatures that predict how patients will respond to treatments and their likely outcomes by linking gene expression or protein levels to clinical data.

Network modeling uses dynamic systems theory and mathematical methods to investigate complex biological systems, showcasing their nonlinear spatiotemporal behaviors. However, generating experimental data needed to define, adjust, and confirm these models can be a lengthy and costly process, and it may not be feasible with existing technologies. Nevertheless, comprehending the complex interactions and changes within the entire system requires mathematical modeling.

Understanding the intricate networks of cancer cells requires more than just intuitive approaches; it demands advanced network models and computational analyses. These models help formulate hypotheses that can be tested in experiments, revealing the mechanisms behind tumor growth and spread. By constructing networks, researchers can pinpoint essential elements of biological processes, which are crucial for exploring how cancer develops and progresses. Both network analysis and modeling deepen our insights into cancer, exposing hidden patterns and unexpected mechanisms while identifying critical areas where knowledge is still lacking. Additionally, these techniques foster the creation of hypotheses that can be validated through experimentation. Network analysis also aids in identifying biomarkers useful for personalized medicine in clinical practice.

2.1. Essential Models for Cancer Systems Biology

Before using systems biology methods, it is important to focus on specific cancer types or subtypes with high clinical relevance and a well-documented molecular pathology. An example of such cancer is breast cancer, which can be stratified into four subtypes based on gene microarray profiles. Interestingly, tumors can vary significantly depending on the source of mammary epithelial cells. Research shows that the gene expression patterns differ based on the specific type of precursor cell, influencing the tumor's characteristics. Thus, understanding the different clinical subtypes of tumors is crucial for precise analysis.

For any given cancer type or subtype, high-quality experimental models are essential. This includes mouse models that represent various cancer stages. For example, progression models for prostate cancer and a range of targeted therapies for specific cancer types or subtypes. The extent to which laboratory animal models (*i.e.*, mice and rats) mimic human cancer types and subtypes plays a crucial role in their ability to reveal molecular mechanisms and guide treatment strategies. Ideally, these animal-based models should accurately reflect patient clinical outcomes for specific cancer types, making them useful for predicting poor chemotherapy responses and aiding in prognosis.

The selected cancer types or subtypes should be backed by well-established cell line models with detailed genomic profiles. A key example is the breast cancer cell line MDA-MB-231, which has several variants that show organ-specific metastasis to the bone, lung, and brain. For these experimental models, researchers can produce extensive molecular datasets, including full gene expression profiles, epigenetic modifications, phosphoproteomics, key cancer-related mutations identified through genome sequencing, and metabolic data. To achieve high-quality data for network building, analysis, and modeling, it is essential for experimental biologists to work closely with computational scientists from the design (first) stage. This collaboration allows for the creation and modeling of tumor-specific networks, which can be incorporated into mathematical frameworks for more advanced simulations.

By leveraging such comprehensive experimental models and interdisciplinary collaboration, systems biology uncovers a valuable understanding of the intricate processes underlying cancer, opening doors to novel treatments and personalized medical approaches.

2.2. Integrating Data for the Development of Cancer Gene Networks

2.2.1. Development of Cancer Gene Networks

The main objective in the biology of cancer systems is to develop dynamic models that represent the biological processes involved in the induction, development, and spread of cancer. These models need to encompass the key functional aspects of cancer biology, with essential processes like the hallmarks of cancer being particularly significant. As our comprehension of cancer biology evolves—especially through the lens of systems biology—we may identify additional hallmarks. For example, inflammation has recently been suggested as a potential new hallmark of cancer. The process of cancer metastasis is influenced by both the inherent characteristics of tumor cells and external factors within the tumor microenvironment, which includes blood vessels and an inflammatory environment filled with immune cells and their secretions that support tumor growth. Tumors exhibit key characteristics linked to cell signaling, including the cell cycle, differentiation, program cell death-related routes, angiogenesis, resistance to inhibitors, invasion and metastasis in various tissues, and inflammation. Integrative analyses of human signaling networks alongside data on cancer driver mutations indicate that cell cycle and program cell death-related pathways are fundamental across various cancer types. Therefore, the investigated networks of cancer- related genes must encompass these biological themes, prioritizing subnetworks that represent them for systems biology research. By isolating these subnetworks, researchers can conduct a more focused analysis of specific cellular processes through network modeling. However, because these processes are interconnected within the cell and tumor microenvironment, it is essential to model their higher-order relationships following comprehensive studies of individual processes.

There are three main strategies for developing cancer gene networks:

• Inference or Reverse Engineering from Genome-Wide Datasets

One strategy involves using datasets that include gene expression profiles and RNA interference (RNAi) knockout phenotype data. For example, researchers constructed a gene regulatory network based on time-course microarray data from a mouse epithelial breast cell line (BRI-JM01) that was induced to undergo epithelial-to-mesenchymal transition (EMT) through TGF-β treatment. In this network, Clusterin, a gene whose expression increases during EMT, exhibited numerous regulatory connections with other genes. Furthermore, inhibiting Clusterin using antibodies was found to suppress TGF-β-induced EMT.

• Extension of Protein Interaction or Signaling Networks

High-throughput experimental techniques, such as protein interaction assays and large-scale ChIP-seq or ChIP-on-chip analyses, can enhance these networks. This method is especially useful for building gene regulatory networks. For instance, researchers developed a P53 regulatory network that identified 98 new direct target genes of P53 using this approach.

• Combining Data from High-Throughput Studies and Literature Databases

This approach leverages manually curated data from the literature to construct current signaling networks. Integrating high-throughput data through computational methods addresses and dissects the complexity of cancer systematically. The quality of data is critical, and the selection of data sources must align with the specific questions posed by the network analysis.

Connecting genes of interest to protein interaction and signaling networks can facilitate the development of cancer-related subnetworks or modules. Alternatively, genes can be incorporated into extensive networks, such as the human signaling network, where the shortest routes between chosen genes can be identified and combined to produce a new network. This approach can involve genes associated with cancer modulation, cancer driver mutations, or other relevant gene groups. Functional genomics data can help pinpoint which shortest paths within the overall network are critical for specific cellular states. For example, signaling networks tailored to particular cancer cell lines can be created by extracting the shortest paths enriched in the gene expression profiles of those lines. Additionally, tumor gene co-expression networks can be developed using data from a variety of tumor types or subtypes. Weighted gene co-expression networks derived from gene microarray profiles of glioblastoma samples serve as a blueprint for harnessing genomic data to uncover key regulatory networks and potential molecular targets in cancer.

Cancer molecular networks can also be constructed by linking genotype and phenotype information. Quigley and colleagues employed this method by interbreeding mouse strains with varying susceptibilities to skin tumors. They combined gene expression profiling with linkage analysis to create a susceptibility network that maps gene expression and regulation in normal skin. Cancer development and metastasis are dynamic processes that unfold over various timescales. Longitudinal high-throughput data can be used to construct networks representing these processes over time. To develop dynamic network models, it is crucial to conceptualize how time is encoded in networks, either as discrete snapshots or continuous sequences of events. Most current experimental systems, however, produce data suitable for discrete-time representation.

All the methods mentioned can construct cancer networks specific to tumor samples or cell lines, thereby clarifying intricate interactions by filtering out extraneous information from the broader network. After constructing these networks, verifying which cancer hallmarks they capture is necessary. If none are found, revisiting the data and network structure processes is essential.

2.2.2. Advancements in Bioinformatics: Unveiling New Frontiers in Systems Biology Data Integration and Analysis

In the field of Systems Biology, integrating data is revolutionizing how we approach bioinformatics analysis. This integration serves as a cornerstone, facilitating the construction, analysis, and modeling of complex networks. A plethora of bioinformatics methodologies and tools have emerged, tailored for handling vast datasets. Statistical tests are crucial for comparing gene expression levels and understanding genetic relationships. These tests help us see how different genes are related to one another. Also, methods for gathering information from research papers and databases, like text mining, are extremely useful. Moreover, there is an increasing use of pattern recognition and machine learning, especially clustering analysis, to help us make sense of complex biological processes.

A noticeable shift in bioinformatics analysis is evident from the traditional focus on individual genes and proteins toward a more holistic perspective. In the nascent stages of microarray analysis, genes were scrutinized in isolation. Presently, the emphasis lies on evaluating the statistical significance of gene expression alterations within gene sets, considering pathways or biological processes like the cell cycle. This paradigm shift has illuminated novel aspects of cancer biology. For example, changes in how proteins within specific networks interact with each other are crucial in the spread of cancer. Noteworthy methodologies like Gene Set Enrichment Analysis (GSEA) have emerged as flag-bearers of this evolving trend. Nonetheless, the field faces formidable challenges in devising bioinformatics methods that are attuned to systems-level intricacies. Chief among these challenges is deciphering the interplay of multivariable factors within biological systems, given the intricate web of interconnected and correlated biological variables.

2.3. Unlocking Complexity: Advancements and Challenges in Network Visualization for Understanding Cancer Molecular Networks

Network visualization serves as a crucial tool in understanding the intricacies of complex systems, particularly in the realm of cancer molecular networks. While mathematical concepts and computational tools have advanced, the human brain still holds a unique capacity for intuition, guiding network analysis in ways algorithms cannot match.

Prior to embarking on network visualization, it is essential to formulate precise questions that extract meaning and implications from these intricate networks. The efficacy of visualization varies with network size, with different objectives depending on scale. Small networks demand a focus on granular graph structures, revealing molecular interactions, while larger networks offer a broader perspective, facilitating judgments on higher-order relationships between subnetworks.

Within large networks, the delineation of network modules and subnetworks is pivotal. Visual analysis helps us understand the internal structures and connections within biological modules. It also allows us to explore the more complex relationships between these modules. By using color-coding based on molecular functions or other important characteristics, we can identify patterns and develop ideas for further investigation. Efforts are underway to evolve network visualization from static representations to dynamic models, incorporating high-throughput and functional data. Dynamic visualization captures the evolving nature of cellular processes, enhancing theoretical understanding and pattern extraction.

Two main methods for visualization have developed: the first combines changes and transitions throughout different stages of development into one network, while the second looks at separate networks for each specific time point. Dynamic network visualization, particularly effective with sparse networks, illustrates the emergence of networks over time by adding and color-coding nodes and relationships as they appear.

Despite advancements in visualization tools like Cytoscape and VisANT, challenges persist, especially in interpreting separate networks at different time points. Comparative network analysis emerges as a powerful approach in such scenarios, offering insights into sequential links between nodes across various time points.

2.4. Navigating the Intricacies of Biological Systems: A Comprehensive Examination of Network Analysis in Cancer Molecular Networks

In the exploration of biological systems, genes and proteins commonly serve as nodes within networks, with the connections between them typically depicted through edges (undirected links) or arcs (directed links).

2.4.1. Different Types of Networks Encode Various Biological Properties

In the realm of cancer molecular networks, a common set of questions arises, yet the diversity of molecular networks allows for addressing different inquiries through distinct network types. These networks vary in their characteristics and relationships encoded, necessitating diverse datasets for their construction.

Gene regulatory networks typically feature short regulatory cascades, reflecting swift regulatory responses within these systems. Hubs within these networks play pivotal roles in stimulus-response and gene coordination, often characterized by rapid transcript decay. These networks help us find important regulators, patterns of gene expression, and groups of genes that play a role in cancer.

Conversely, human signaling networks exhibit sparser connections, with longer regulatory cascades, emphasizing logical regulatory relationships. These networks are essential for identifying genes that cause cancer and understanding the rules that control cancer processes.

- **Signaling networks** show how signals are transmitted and how regulatory hierarchies work, while **gene regulatory networks** reveal the order of regulation.

- The biological meanings of connections in protein interaction networks are less clear, although these networks show unique evolutionary traits.

- In signaling networks, key regulatory actions happen through protein modifications instead of changes in gene expression. Therefore, tracking phosphorylation dynamics is vital for understanding these networks.

Network motifs in protein interaction networks represent protein complexes. In signaling networks, they act as information processing units, while in gene regulatory networks, they indicate regulatory loops.

- Protein interaction networks are densely connected and rich in modules, providing insights into cancer development and progression, as well as potential biomarkers.

- In contrast, signaling networks are more sparse and encode logical regulatory codes, whereas protein interaction networks are dense but lack regulatory logic.

- Gene regulatory networks combine regulatory logic with genes that act as "workers."

- It is also important to consider post-transcriptional and post-translational regulations when building, analyzing, and modeling these networks.

- **Ubiquitination**, which occurs in the early stages of signaling networks, and **microRNA regulation**, which happens in the later stages, offer feedback mechanisms in gene regulatory networks.

- **Hub** transcription factors often regulate multiple microRNAs, highlighting their regulatory significance.

2.4.2. Exploring Biological Systems through Network Analysis

In the realm of cancer research, the principles of evolution are central. Just like the laws of physics and chemistry, we can discover and model the basic principles that govern cancer biology. To effectively communicate the mechanisms at play, we need to model the "core design principles" in biology. Abstraction is essential for revealing these design principles. It helps us analyze data from different angles and allows us to extract valuable insights. Graph theory provides a means to represent biological relationships, analyze information, and extract insights, with evidence showing that biological perceptions are reflected in the characteristics of the networks. Therefore, studying the characteristics of integrated networks—like combining signaling networks with cancer-related high-throughput omics data—can provide new biological insights.

Important ideas in network analysis include non-linear approaches and viewing networks from different perspectives. Non-linear thinking often leads to the discovery of emergent biological properties, whereas linear thinking typically yields predictable results. Therefore, network analysis, with its propensity for unexpected outcomes, can lead to more exciting discoveries.

Network properties span from local to global scales, encompassing features like single nodes or edges, network bottlenecks, motifs, modules, and network-wide characteristics. Even small changes in a few connected nodes can have a big impact, highlighting the complex links between local and global properties. For example, changes in the gene co-expression of protein network modules can contribute to the spread of breast cancer.

Different methods of network analysis can be employed based on specific research questions. Global network features can address questions related to cancer signaling functionality, genetic and epigenetic impacts on tumor progression, identification of central players (*e.g.*, cancer genes), and functional targeting of cancer hallmarks. Local network features can help us understand the presence of cancer-related mutated genes and methylated genes within signaling loops, providing insights into how oncogenes function and contribute to tumor formation.

Additionally, network analysis is a valuable tool for generating hypotheses that can be tested. For example, finding a high concentration of ubiquitin-mediated proteins in positive feedback loops of the human signaling network led to the idea that the ubiquitination process might be more active in tumor cells, preventing apoptosis, which is a crucial step in cancer signaling. This hypothesis was supported by evidence from microarray data analysis of tumor and normal samples.

Overall, network biology provides valuable insights into cancer biology, offering a departure from traditional linear approaches and paving the way for novel discoveries and hypotheses.

2.4.3. Analyzing Network Dynamics for Diagnosis and Prognosis through Gene Markers

The natural genetic differences and somatic mutations found in human populations, including tumors from the same person, lead to the significant diversity seen in tumors. This variation means that patients with the same type of cancer can have different responses to treatments, highlighting the importance of personalized medicine to identify molecular markers that predict how well drugs will work.

Another challenge in the treatment of cancer lies in predicting which patients may benefit from additional therapy post-surgery. Currently, predicting the necessity of additional therapy after surgery remains challenging, as it's difficult to determine which patients will respond well to surgery alone and which may require additional chemotherapy. Thus, identifying genes that can serve as prognostic tools for survival after cancer diagnosis is crucial for guiding treatment decisions and achieving optimal outcomes.

Despite extensive efforts over the past decade, robust molecular markers for cancer, particularly breast cancer, remain elusive. The effectiveness of current breast cancer biomarkers often diminishes when applied to different sets of tumor samples, highlighting the need for more accurate and reliable markers.

A systems biology approach shows potential for finding markers that explain how patients respond to drug treatments by analyzing the dynamics of the entire system. Integrating cellular networks with genetic, epigenetic, and protein modification data alongside clinical information can revolutionize cancer management strategies.

Recent studies show that using a network biology approach is effective in finding better markers for predicting cancer outcomes. By placing tumor-related genes onto a human protein interaction map, researchers can identify smaller networks that serve as cancer biomarkers. These smaller networks, subnetwork, markers exhibit higher reproducibility and accuracy in classifying metastatic versus non-metastatic tumors compared to individual marker genes. Additionally, changes in the co-expression of genes within protein network modules are linked to cancer spreading, showing how protein interaction modules are reorganized during metastasis. The markers of the network modules have proven better at predicting survival in breast cancer patients compared to markers based on gene clustering, highlighting the importance of viewing genes as part of interacting modules. Using network biology to discover drug response markers shows great potential for improving cancer treatment strategies in the future.

2.5. Dynamic Network Modeling: Deciphering the Information Processing Machinery of Cancer Cells

Biology is undergoing a significant shift toward understanding system dynamics, especially for studying interconnected systems, information processing like cell signaling, and key computational and functional units. Organisms are now often seen as advanced processors and manipulators of information.

Dynamic network modeling, based on principles from dynamical systems theory, is instrumental in decoding the information processing mechanisms within cancer cells and testing hypotheses about their functioning. This modeling approach quantitatively captures how cancer cells behave by tracking changes in gene activity or enzyme levels. Network modeling offers conceptual and computational tools that support dry-lab experiments, allowing simulation-based research within a clear, quantitative framework. In the long run, it could replace many expensive and time-consuming wet-lab experiments, leading to major breakthroughs in cancer biology through the growing field of systems biology.

Two fundamental approaches to dynamic network modeling, qualitative and quantitative, are commonly employed. Qualitative modeling categorizes network node states into finite values (*e.g.*, ON/OFF), while quantitative modeling spans a wider range of values, taking into account probabilistic, deterministic, or random factors.

The distinguished models of the dynamic network consist of three key factors: the cellular network, initial node states, and transfer functions describing node dependencies. These models can adopt either continuous or discrete representations, deterministic or stochastic, leading to four modeling methods.

Continuous stochastic models theoretically offer more accurate descriptions but demand high-quality experimental data. Conversely, discrete deterministic models, such as Boolean models, offer simpler representations suitable for current experimental data but yield more macro-scale and less quantitative predictions.

In summary, systems biology is revolutionizing cancer biology, yet high-quality quantitative data and new modeling methods are essential. Moreover, there is a critical need to train more investigators to think in terms of networks rather than linearly, fostering advancements in dynamic network modeling and its application to cancer research.

Vocabulary (Index of Terms)

Apoptosis [ˌæpəpˈtoʊsɪs]	Apoptosis is a type of programmed cell death that takes place in multicellular organisms. It is an important process for maintaining the balance of cell proliferation and death, eliminating damaged or unnecessary cells.
Bioinformatics [ˌbaɪoʊɪnfərˈmætɪks]	Bioinformatics involves the application of computational techniques to analyze and interpret biological data. It plays

a critical role in managing and understanding large-scale genomic and proteomic data.

Cancer Driver-Mutating Genes [ˈkænsər ˈdraɪvər mjuːˈteɪtɪŋ dʒiːnz]

These genes harbor mutations that contribute to the initiation and progression of cancer. Identifying and studying these genes is essential for understanding cancer mechanisms and developing targeted therapies.

Chemotaxis [ˌkiːməˈtæksɪs]

The directed movement of an organism or cell in response to an external chemical stimulus. This movement is typically aimed at finding or avoiding specific substances in the environment.

Cytokine [ˈsaɪtəˌkaɪn]

Cytokines are small proteins released by cells, particularly those in the immune system, that act as signaling molecules to regulate immunity, inflammation, and hematopoiesis.

Epigenetics [ˌɛpɪdʒəˈnɛtɪks]

Epigenetics refers to the study of heritable changes in gene function that do not involve changes in the DNA sequence. These changes can affect gene activity and expression.

Gene Expression Profiling [dʒiːn ɪkˈsprɛʃən ˈproʊfaɪlɪŋ]

This technique measures the activity (expression) of thousands of genes at once to create a global picture of cellular function. It is used to understand gene function, disease mechanisms, and response to treatments.

Gene Regulatory Networks [dʒiːn ˈrɛgjʊlətɔri ˈnɛtwɜrks]

These networks depict the interactions between different genes and how they regulate each other's expression. They are crucial for understanding complex biological processes and disease states.

Hallmarks of Cancer [ˈhɔlmɑrks ʌv ˈkænsər]

The hallmarks of cancer are characteristics that are commonly observed in cancerous cells. These processes encompass maintaining continuous signaling for cell growth, avoiding growth inhibitors, resisting cell death, achieving unlimited replication, promoting the formation of new blood vessels, and facilitating invasion and spread to other tissues.

Metastasis [məˈtæstəsɪs]

Metastasis is the process by which cancer cells spread from the place where they first formed to other parts of the body. This spread can occur through the blood or lymphatic system.

Microarray [ˈmaɪkroʊˌreɪ]	A microarray is a laboratory tool used to detect the expression of thousands of genes simultaneously. It consists of a small surface onto which DNA molecules are fixed in a grid-like manner.
Network Modules [ˈnɛtwɜrk ˈmɑdʒʊlz]	Network modules are subnetworks within a larger network that consists of highly interconnected components. These modules often represent distinct functional units within biological systems.
Oncogene [ˈɒŋkəˌdʒiːn]	An oncogene is a gene that can lead to cancer development. In tumor cells, these genes are frequently mutated or overexpressed, which helps cancer cells grow and survive.
Phosphoproteomics [ˌfɒsfoʊˌproʊtiˈoʊmɪks]	Phosphoproteomics is the study of phosphorylated proteins, which are proteins modified by the addition of a phosphate group. This field helps in understanding signaling pathways and cellular processes in cancer.
Proteomics [ˌproʊtiˈoʊmɪks]	Proteomics involves the comprehensive analysis of proteins, focusing on their structures and functions. It is crucial for understanding cellular processes and disease mechanisms.
Systems Biology [ˈsɪstəmz baɪˈɒlədʒi]	Systems biology is an interdisciplinary field that focuses on complex interactions within biological systems. It aims to understand and model these systems comprehensively by integrating various types of data.
Ubiquitination [juːˌbɪkwɪtɪˈneɪʃən]	Ubiquitination is a process in which a ubiquitin-protein is attached to a substrate protein. This modification can signal the protein's degradation or alter its cellular location and function.

COMPREHENSION TASKS

I. Key Terms: Matching

Instructions: Match each term from the left column with the corresponding description from the right column. Write the letter of the correct description next to the term's number. Each description will only match with one term. Good luck!

1. Angiogenesis	**A.** Addition of chemical groups to histone proteins affecting gene expression.
2. Biomarker	**B.** Process by which cancer forms.
3. Chromatin	**C.** Analysis of RNA transcripts to study gene expression.
4. DNA Methylation	**D.** Short sequences of RNA regulating gene expression post-transcriptionally.
5. Enzyme-Linked Immunosorbent Assay (ELISA)	**E.** Protective caps at the ends of chromosomes.
6. Genomic Instability	**F.** Uncontrolled division of cells leading to tumor development.
7. Histone Modification	**G.** Technique to amplify DNA sequences.
8. Immunohistochemistry	**H.** Use of antibodies to detect specific antigens in tissues.
9. Kinase	**I.** Measure the stability and integrity of the genome.
10. MicroRNA	**J.** Enzyme adding phosphate groups to proteins.
11. Oncogenesis	**K.** Formation of new blood vessels.
12. PCR (Polymerase Chain Reaction)	**L.** Laboratory method for detecting and quantifying proteins.
13. Signal Transduction	**M.** Measurement and interpretation of biological markers.
14. Telomere	**N.** Complex of DNA and protein in the nucleus.
15. Transcriptomics	**O.** Chemical modification of DNA affecting gene expression.

II. Cell Biology True or False Quiz:
Determine whether the following statements are true or false based on your understanding of cell biology concepts. Write "True" if the statement is correct and "False" if it is not. Good luck!

1. ----- Angiogenesis refers to the formation of new blood vessels.

2. ----- Biomarkers are used to measure and interpret the stability of the genome.

3. ----- Chromatin is composed of DNA and proteins found in the cytoplasm.

4. ----- DNA methylation can affect gene expression by adding chemical groups to DNA.

5. ----- ELISA is a technique used to amplify DNA sequences.

6. ----- Genomic instability is a hallmark of many cancers.

7. ----- Histone modification involves the addition of chemical groups to DNA.

8. ----- Immunohistochemistry uses antibodies to detect specific proteins in tissue samples.

9. ----- Kinases are enzymes that remove phosphate groups from proteins.

10. ----- MicroRNAs regulate gene expression by interacting with mRNA.

11. ----- Oncogenesis refers to the transformation of normal cells into cancer cells.

12. ----- PCR is a technique used to study protein-protein interactions.

13. ----- Signal transduction pathways are crucial for transmitting signals within a cell.

14. ----- Telomeres are chromosome ends that protect against degradation.

15. ----- Transcriptomics is the study of all RNA transcripts generated by the genome.

Cell Biology Completion Exercise: Fill in the blanks with the appropriate term or phrase to complete each sentence related to cell biology. Write your answers in the space provided. Good luck!

1. Angiogenesis is the process of forming new _____.

2. Biomarkers can be used to detect and measure _____ changes in cells.

3. Chromatin is a complex of DNA and _____ that forms chromosomes within the nucleus.

4. DNA methylation typically acts to _____ gene expression by adding methyl groups to DNA.

5. ELISA is a laboratory technique used to detect and quantify _____ in a sample.

6. Genomic instability often leads to increased rates of _____ and is a common feature of cancer.

7. Histone modification involves the addition or removal of _____ groups to histone proteins, which influences chromatin structure and gene expression.

8. Immunohistochemistry is a method for detecting specific _____ in tissue samples using antibodies.

9. Kinases are enzymes that catalyze the addition of _____ groups to proteins, often regulating their activity.

10. MicroRNAs are tiny non-coding RNA molecules that control gene expression by attaching to specific _____ and preventing their translation.

III. Multiple Choice: Choose the best answer for each question by circling the corresponding letter.

1. Which process involves the formation of new blood vessels?

a. Hematopoiesis

b. Angiogenesis

c. Apoptosis

d. Phagocytosis

e. None of the above

2. Biomarkers are used primarily to:

a. Destroy pathogens

b. Enhance cell growth

c. Detect and measure biological changes

d. Repair DNA

e. None of the above

3. The complex of DNA and proteins that forms chromosomes within the nucleus is called:

a. Chromatin

b. Nucleolus

c. Ribosome

d. Centromere

e. None of the above

4. DNA methylation typically:

a. Increases gene expression

b. Suppresses gene expression

c. Repairs damaged DNA

d. Increases protein synthesis

e. None of the above

5. ELISA is a laboratory technique used to:

a. Sequence DNA

b. Detect and quantify proteins

c. Visualize cell structures

d. Measure cell motility

e. None of the above

6. Genomic instability often leads to:

a. Decreased cell division

b. Increased rates of mutations

c. Stabilized gene expression

d. Decreased protein synthesis

e. None of the above

7. Histone modification affects:

a. Chromatin structure and gene expression

b. Mitochondrial function

c. Membrane transport

d. Ribosome assembly

e. None of the above

8. Immunohistochemistry is used to detect specific:

a. Lipids

b. Carbohydrates

c. Nucleotides

d. Proteins

e. None of the above

9. Kinases are enzymes that:

a. Remove phosphate groups from proteins

b. Add phosphate groups to proteins

c. Break down DNA

d. Synthesize lipids

e. None of the above

10. MicroRNAs regulate gene expression by binding to:

a. DNA

b. mRNA

c. Proteins

d. Ribosomes

e. None of the above

11. Which of the following is uncommon in cancer?

a. Genomic instability

b. Increased mutation rates

c. Stable gene expression

d. Uncontrolled cell growth

e. None of the above

12. Angiogenesis is critical for:

a. Wound healing

b. Immune response

c. Neuron function

d. Muscle contraction

e. None of the above

13. Biomarkers can be found in:

a. Blood

b. Urine

c. Tissue samples

d. All of the above

e. None of the above

14. Chromatin is primarily composed of:

a. DNA and lipids

b. RNA and proteins

c. DNA and proteins

d. Carbohydrates and lipids

e. None of the above

15. DNA methylation involves the addition of methyl groups to:

a. RNA

b. Proteins

c. DNA

d. Lipids

e. None of the above

16. The main function of ELISA is to:

a. Sequence DNA

b. Amplify RNA

c. Detect and quantify antigens

d. Visualize cell structures

e. None of the above

17. Genomic instability can result from:

a. DNA replication errors

b. Environmental factors

c. Both a and b

d. Neither a nor b

e. None of the above

18. Histone modifications can include:

a. Methylation

b. Acetylation

c. Phosphorylation

d. All of the above

e. None of the above

19. Immunohistochemistry utilizes:

a. Enzymes

b. Antibodies

c. Fluorescent dyes

d. Radioactive isotopes

e. None of the above

20. MicroRNAs typically:

a. Enhance translation

b. Inhibit translation

c. Enhance DNA replication

d. Inhibit DNA replication

e. None of the above

READING UNDERSTANDING

Diagnosis of Diseases in Livestock

A cow lies in the middle of the pasture after the herd has moved on to be milked. The owner examines the cow closely and tries to get her to stand. She makes a feeble attempt but is either too weak or too distressed to succeed. Is the cow injured? Has she ingested something toxic? Is she suffering from a severe systemic infection? Could there be a functional issue in her nervous, digestive, or urogenital system? The process of identifying the disease is known as diagnosis. This involves systematically gathering facts and objectively evaluating them while considering known diseases and their symptoms.

I. *Importance of Diagnosis*

Why is making a diagnosis crucial? In the case of the downer cow, diagnosing the disease is essential for the protection of the herd. If the disease is contagious, it must be identified and contained to prevent further losses. Secondly, accurate diagnosis is critical for determining the correct treatment approach to restore the cow's health.

II. *Diagnostic Approach*

Experienced diagnosticians avoid jumping to conclusions, even when symptoms seem to clearly point to a specific disease. Hasty evaluations are prone to errors. Diagnosticians collect information by questioning the owner and examining records about the herd's health, production, reproduction status, nutrition, past illnesses and treatments, vaccinations, recent behaviors, and any signs observed by the owner. A thorough examination of the cow follows, assessing her overall appearance, condition, and mental state. This physical examination includes measuring body temperature, respiratory rate and quality, and pulse rate and quality.

- **Body Temperature:** A fever may indicate a systemic infection or high environmental temperature. A decrease might suggest intoxication.

- **Respiratory Rate:** Increased respiration could signal excitement, fever, lung infection, or anemia.

- **Pulse Rate:** Changes in pulse can be affected by pain, fever, blood volume, and circulatory changes.

The absence of changes from normal values is just as important as the presence of changes. The examination continues with a detailed evaluation of each system—circulatory, respiratory, digestive, nervous, musculoskeletal, urogenital—either separately or concurrently. If the diagnosis remains uncertain after a complete physical examination, laboratory analysis of tissues, blood, secretions, or excretions may be necessary. Radiographs and other diagnostic tools can also be useful but may be inconvenient in a pasture setting.

III. Post-Diagnosis Actions

Once a diagnosis is made, appropriate measures can be taken to control or prevent the disease from spreading and to treat the affected cow.

- **Metabolic Disease:** Adjust nutritional intake.
- **Infectious Disease:** Implement steps to prevent other cattle from coming into contact with the infected cow.
- **Toxicity:** Identify and remove the source of the toxin, whether it is a plant or chemical.
- **Tumors or Anomalies:** These generally do not pose a threat to the herd.
- **Traumatic Injuries:** Identify and eliminate the cause of the injury.

IV. Diagnosis of Deceased Animals

Even if the cow has already died, diagnosis is still crucial to protect the herd. The urgency may be even greater, as the cause of death might pose a threat to the remaining animals. The diagnostic process in such cases involves a postmortem examination or necropsy to gather facts from the carcass.

V. Preventing Disease Outbreaks

Since farm animals are often kept in close proximity, disease outbreaks can spread quickly and cause severe losses. Immediate and proactive steps are necessary to eliminate or mitigate the cause and protect susceptible animals. Ensuring that no sick or dead animal goes undiagnosed can significantly reduce losses. Veterinarians are trained in the art of diagnosis, with their education and experience aimed at developing a deep understanding of disease signs and processes.

1. What is the primary goal of making a diagnosis in livestock?

A. To improve the cow's milk production

B. To identify the cause of the disease and prevent its spread

C. To determine the best diet for the herd

D. To evaluate the physical fitness of the cow

2. Why is it important to avoid jumping to conclusions when diagnosing a cow's illness?

A. To prevent unnecessary treatments

B. To save time and resources

C. Because hasty evaluations are prone to errors

D. To ensure the cow gets enough rest

3. Which of the following is NOT a method used in the physical examination of a cow?

A. Measuring body temperature

B. Assessing the respiratory rate

C. Evaluating mental attitude

D. Performing a blood transfusion

4. What could a decrease in body temperature indicate in a cow?

A. Systemic infection

B. High environmental temperature

C. Intoxication

D. Excitement

5. If the exact disease remains unclear after a physical examination, what might a veterinarian do next?

A. Administer a general antibiotic

B. Perform a laboratory analysis of tissues and blood

C. Increase the cow's food intake

D. Separate the cow from the herd

6. What should be done if the disease affecting a cow is found to be infectious?

A. Increase the cow's water intake

B. Isolate the infected cow to prevent contact with others

C. Change the cow's diet

D. Ignore and monitor the situation

7. Which diagnostic tool might be inconvenient to use in an open pasture?

A. Body temperature measurement

B. Respiratory rate assessment

C. Pulse rate check

D. Radiographs

8. Why is diagnosis still crucial even if the cow is already dead?

A. To determine the exact time of death

B. To assess the overall health of the carcass

C. To protect the rest of the herd from the same fate

D. To decide if the meat is still consumable

9. What is a necropsy?

A. A type of blood test

B. A postmortem examination to determine the cause of death

C. An evaluation of the cow's mental state

D. A vaccination process for the herd

10. How can losses from disease be minimized in a herd of farm animals?

A. By increasing the feeding frequency

B. By making sure all sick or dead animals are diagnosed

C. By reducing the herd size

D. By limiting outdoor grazing

CHAPTER 11

The Evolution and Future Prospects of the ISI Web of Knowledge Platform

Abstract: The ISI Web of Knowledge, evolving from Eugene Garfield's pioneering citation indexes, represents a sophisticated web-based platform central to modern scholarly research. This paper explores its transformative journey from print citation indexes to a dynamic, integrated digital environment. Initially, citation indexes such as the Science Citation Index® facilitated scholarly communication by linking references and tracking research impact. The transition to the ISI Web of Knowledge marked a significant advancement, incorporating advanced technologies for seamless access to multidisciplinary content. This platform features enhanced search capabilities, including ISI CrossSearch and ISI eSearch, and robust linking systems like ISI Links and RoboLinks. By integrating context-sensitive linking *via* SFX, the ISI Web of Knowledge supports comprehensive access to diverse information sources, thus addressing challenges of information overload and accessibility. Future developments promise further expansion of content and capabilities, solidifying the platform's role as a pivotal resource in the landscape of academic research and bibliometrics.

Keywords: Bibliometrics, Citation indexes, Information access, Scholarly communication, Web-based research platforms.

UNDERSTANDING THE CHAPTER'S FOCUS: A ROADMAP

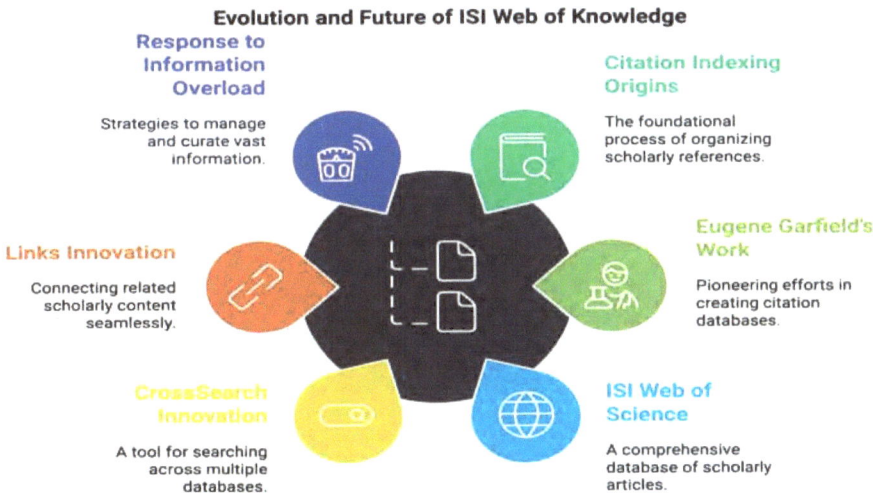

Evolution and Future of ISI Web of Knowledge

Response to Information Overload — Strategies to manage and curate vast information.

Citation Indexing Origins — The foundational process of organizing scholarly references.

Links Innovation — Connecting related scholarly content seamlessly.

Eugene Garfield's Work — Pioneering efforts in creating citation databases.

CrossSearch Innovation — A tool for searching across multiple databases.

ISI Web of Science — A comprehensive database of scholarly articles.

Mohammad Mehdi Ommati

1. OVERVIEW

In 1955, Eugene Garfield reported a groundbreaking article in Science called "Citation Indexes for Science: A New Dimension in Documentation through Association of Ideas." At present, Dr. Garfield is recognized as a pioneer in the field of bibliometrics, which is the study of how information is cited and used in research. His original idea of building a citation database that covers multiple disciplines has greatly changed over the years, now using advanced technologies that were not even possible 45 years ago. This evolution has transitioned citation data from print to electronic formats, eventually creating a web-based environment characterized by hypernavigation, intuitive linking, and sophisticated contextual interactions.

This paper will explore how citation indexes evolved into the ISI Web of Knowledge, an integrated resource platform widely used by research libraries around the world. It will start by discussing Dr. Garfield's important contributions to bibliometrics, which formed the foundation for the ISI® platform. The paper will then focus on three key aspects: first, the platform's ability to offer access to a diverse range of multidisciplinary content, which is vital for modern research; second, the innovative query engine for knowledge management that allows users to search across different types of content; and third, ISI Links, the main system that provides confirmation, access, and routing, which supports the platform's extensive linking features.

2. THE ROLE OF CITED REFERENCES IN SHAPING SCIENTO-METRICS

A cited reference is a way for an author to acknowledge the influence of other scholars in their work. Citation indexes take this concept further by showcasing a wide range of influences from different authors and works. One major benefit of citation indexing is that it helps researchers find relevant articles across various fields without needing to know specific jargon—the citation itself can serve as a useful search term.

Citation indexes allow users to discover related papers, see how their research is being used by others, and find connections between different research topics that may seem unrelated at first. This makes citation indexes an essential tool for both researchers and those studying the history of science, as they help track the development of scientific knowledge and reveal links between different areas of study.

Dr. Garfield understood these relationships well. In 1961, ISI received funding from the NIH to create the "Genetics Citation Index," which was a forerunner to the Science Citation Index®. Garfield pointed out that the citations found at the end of a scientific paper do more than just give credit to other researchers; when these citations are indexed and included in a database, they can help track and link important research developments over time.

The Science Citation Index was first released in 1964 as a five-volume set that included 613 journals and 1.4 million citations. While this was an exciting new resource for accessing scientific research, finding information was often slow and cumbersome because it was in print format. Two years later, the citation data became available on magnetic tapes, enabling institutions with the right technical skills to automate the handling of citation data.

As technology improved, so did ISI's offerings. The Science Citation Index was expanded in 1972 with the introduction of the Social Sciences Citation Index® and in 1978 with the Arts & Humanities Citation Index®, which eventually moved to online platforms and later to CD-ROMs. In recent years, the time it takes for new information to become part of the global knowledge base has greatly decreased. The expansion of electronic access to scholarly literature has played a crucial role in this acceleration. While the publication and distribution of printed journals and indexes once meant that communication among researchers could take months or even years, current technology has reduced this cycle to mere days. The widespread availability of personal computers, the Internet, and electronic journals has dramatically enhanced global access to scientific information.

For many years, researchers in bibliometrics and scientometrics have depended on citations, utilizing ISI citation indexes as their main source of data for analysis. However, it was not until 1997, when these citation indexes were transitioned to a web-based platform known as the ISI Web of Science®, that researchers began to actively participate in searching for cited references.

2.1. How 'Citation Indexes' Transformed into the 'ISI Web of Science'

The development of hypertext links and web browsers made it possible to create the Web of Science, a user-friendly and easy-to-understand tool for tracking citations. Dr. Garfield's innovative idea of organizing scientific literature was far ahead of its time, and he had to wait over 40 years for technology to catch up. Standardizing how references were recorded was a crucial first step, but the ISI citation indexes did not fully evolve into the Web of Science until the development of internet technologies, like advanced linking and search functions.

The introduction of hypernavigation changed both general and citation searches completely. When ISI's citation database became available online, it greatly improved basic bibliographic searches, allowing researchers to explore historical and interdisciplinary studies more easily. Using the "Cited References" link, researchers could trace back to the original studies that influenced a paper, while the "Times Cited" link showed how that paper impacted future research. The "Related Records®" feature also helped users find other relevant papers quickly, uncovering interdisciplinary connections.

The effectiveness of the linking and citation searching capabilities largely hinges on the accuracy of reference capture—the meticulous process of recording bibliographic entries. ISI has always followed strict standardization processes, using a unique algorithm to create a specific key for each document. Every year, 23 million references are checked against a detailed reference system, standardized, and prepared for inclusion in the ISI Web of Science. For more than 40 years, ISI has maintained high bibliographic standards, which made it possible to move from print citation indexes to the dynamic, web-based system we have today.

Transitioning citation indexes to a web platform was a significant milestone for citation searching and analysis. However, it was the combination of web-based search and linking technologies, along with the expansion of online content, that changed the field. The ISI Web of Knowledge platform now brings together research-focused content, provides access to full-text documents, offers reliable links, and uses advanced search engines, creating a smooth and comprehensive research environment for scholars and scientists around the world. This transformation has not only expedited the dissemination of scientific knowledge but also enhanced the ability to discover and explore connections across diverse fields of study.

2.2. The ISI Web of Knowledge: A Comprehensive Research Tool

2.2.1. Introduction to the ISI Web of Knowledge

This is a sophisticated, web-based platform specifically aimed to facilitate scientific and scholarly research across various sectors, including academic, corporate, governmental, and non-profit organizations. This platform's core concept is elegantly straightforward: It combines high-quality, carefully reviewed content with essential tools for using, analyzing, and managing that information. The overarching objective of the platform is to enable researchers and scholars to effortlessly access and navigate the wealth of information they require daily, directly from their desktops, seamlessly integrating it into their research workflows.

2.2.2. Enhancing Research Through Seamless Access

To encourage intellectual exploration and accelerate discoveries, institutions should enable researchers to pursue their studies with as few external barriers as possible, such as environmental constraints or resource limitations. Research endeavors can be significantly disrupted if available information resources are disjointed, challenging to access, or if there are restrictions on their use in terms of time and location. The ISI Web of Knowledge mitigates these issues by providing a comprehensive information solution that supports and accelerates scholarly research rather than inadvertently hindering it. To fully grasp how this platform achieves such support, it is crucial to examine both the extensive content it offers and the advanced tools it integrates, particularly those related to searching and linking.

2.2.3. Comprehensive Content and Advanced Tools

The platform's design ensures that high-quality content is readily available, facilitating a robust research environment. Its advanced tools are tailored to enhance the search experience and provide intuitive linking capabilities, which are vital for thorough and efficient scholarly investigation. By reviewing these aspects, one can understand the platform's potential to significantly impact the research process, making it more fluid and less prone to disruptions, thereby promoting a more effective and streamlined approach to academic and scientific exploration.

3. THE VALUE OF MULTIDISCIPLINARY CONTENT IN ISI WEB OF KNOWLEDGE

3.1. Unified Access to Diverse Information

The ISI Web of Knowledge platform provides a complete and integrated environment where researchers can easily search for and access various types of information, including journal articles, conference proceedings, patents, chemical reactions and compounds, and web content at both the site and document levels. This multidisciplinary platform allows scholars to explore beyond specific subjects, which is increasingly important as research becomes more interdisciplinary.

3.2. The Evolution of Global Research Accessibility

The growth of email, the popularity of the internet, and the rise of electronic journals have greatly changed how researchers access information around the world. These tools help scholars easily find, read, and use research from different fields. Because of this, having access to a mix of subjects is very important for major research institutions. To keep up with new research and trends, scientists and researchers need a clear, global view of all the published literature.

3.3. Challenges of Information Overload

The rapid expansion of scientific information presents a paradox. While it offers unprecedented opportunities for discovery, it also imposes significant demands on researchers to keep up with trends and key findings within their fields. The challenge has shifted from merely finding information to finding relevant, high-quality information. In today's world, the high-quality, thoroughly checked research information provided by ISI is extremely important. With so much information available, having reliable and well-evaluated content helps researchers find trustworthy data to support their work. This reliable research plays a crucial role in advancing knowledge and understanding across various fields.

3.4. Streamlined Access to Relevant Information

The ISI Web of Knowledge platform offers a unified access point to important scholarly information. Journals included in the ISI database undergo a thorough review by editorial experts. Conference proceedings are selected to maintain a multidisciplinary and international scope. Specialty editors at Derwent Information™ carefully examine patent information from 40 different sources. They then add detailed descriptions to each record, making the information easier to understand for both experts in patents and those who are not. Senior chemists review chemical information, chosen using standards created in collaboration with leading pharmaceutical companies and universities. They ensure that new synthetic methods are included, along with thorough indexing for easy reference.

3.5. Evaluated Web Content for Scholarly Use

The importance of providing access to vetted, relevant resources extends to web content. The ISI Web of Knowledge platform offers a curated selection of websites and individual documents. Searching for reliable academic resources online can be difficult, but ISI eases this process by thoroughly evaluating websites. They only include those that meet high standards for reliability, accuracy, relevance, design, usability, content depth, target audience, and writing quality.

3.6. Expanding Content Offerings through Strategic Partnerships

ISI's mission is to offer reliable content for researchers and scholars. The ISI Web of Knowledge platform adds to this by including information from other sources. Right now, it offers BIOSIS Previews® for biology, and soon it will add CAB Abstracts® for agriculture and INSPEC® for engineering, making it even more useful across different fields.

4. ISI WEB OF KNOWLEDGE: LEVERAGING INNOVATIVE TOOLS THROUGH ADVANCED TECHNOLOGIES

4.1. Enhancing Research Capabilities with Advanced Tools

A content source is valuable not just for its relevance and coverage, but also for the tools it offers to help users easily access, analyze, and organize the information. In the last half-century, researchers have extensively utilized various tools within individual ISI resources, which have become integral to the research process. These tools encompass cited reference searches using ISI Web of Science, alert services from ISI Current Contents Connect, citation metrics provided by ISI Journal Citation Reports and ISI Essential Science Indicators, and personal bibliography management through EndNote, ProCite, and Reference Manager. The platform of the ISI Web of Knowledge has combined these resources into a single, user-friendly environment, enhancing their functionality. Furthermore, ISI's modular infrastructure, developed over the last decade, enables the smooth addition of new technologies as they become available. One of the first innovations incorporated is a new search tool from Smartlogik Group PLC that utilizes a probabilistic model to explore both structured and unstructured information within the platform. This advancement allows for cross-content concept searching with ISI CrossSearch and provides an improved search capability for web documents through ISI eSearch.

4.2. ISI CrossSearch and eSearch: Unifying Access to Diverse Research Resources

ISI CrossSearch offers researchers a single platform to search four types of content: journal articles from ISI Web of Science, conference papers from ISI Proceedings, patents from Derwent Innovations Index, and bioscientific information from BIOSIS Previews. In its first phase, CrossSearch acts as a gateway to the research environment, featuring a "concept-based" search box that uses probabilistic methods to deliver results ranked by relevance. Users can perform a general search across all content, view a combined summary of their results without duplicates, and then choose specific sources to access full records and special search features, such as cited reference searches in ISI Web of Science. CrossSearch streamlines access by providing a single URL for multiple content sources, allowing users to work across these sources in one session. Additionally, it employs dynamic load balancing and seamless server switching to ensure optimal performance.

ISI eSearch uses advanced indexing and retrieval technology from muscatdiscovery to further integrate the platform. This feature enables users to access both journal articles and web documents from ISI-reviewed websites through a single search in CC Connect. By utilizing Boolean terms from a researcher's CC Connect query, the system uses a front-end query processor developed at ISI to conduct simultaneous sub-queries: one to define the results set and another to enhance relevance through probabilistic weighting. This seamless integration helps researchers find useful web resources as part of a typical journal article search without having to learn a new interface. Just as CrossSearch connects journals, conference proceedings, and patent information, Current Contents eSearch breaks down the barriers between structured content, like scholarly journals, and unstructured web content, which includes PDF, PostScript, HTML, and plain text files, often known as "grey literature."

4.3. Future Directions for ISI Web of Knowledge

By combining ISI CrossSearch and Current Contents eSearch, the ISI Web of Knowledge platform is poised to become a robust research portal. Plans for future enhancements include:

- Increasing the range of content sources to encompass additional specialized databases, such as CAB Abstracts for agricultural data and INSPEC for engineering resources.

- Creating a cohesive access point for both proprietary and external materials within academic institutions.

- Incorporating features currently available in individual resources, like alert notifications, across all platform areas.

- Allowing customization options and institutional branding for a personalized portal experience.

Additionally, the platform will regularly explore new technologies for access and linking that can be integrated into its infrastructure. Guidance from initiatives such as the Open Archives Initiative will shape future updates, ensuring that ISI Web of Knowledge remains a leader in promoting innovative and effective scholarly research.

5. ISI LINKS: AN INTEGRATED LINKING SOLUTION

A key component of the ISI Web of Knowledge platform is ISI Links, which significantly enhances the user experience by facilitating connections to a wide range of resources. While the platform's modular structure and advanced search capabilities are notable features, ISI Links is what truly empowers users to access information seamlessly. So, what does ISI Links do? Essentially, it acts as an intelligent system designed to help you quickly locate and connect to various types of academic content. Whether you're searching for journal articles, conference proceedings, or other scholarly materials, ISI Links simplifies the process, ensuring you can navigate effortlessly between different resources.

In recent years, ISI has focused on building a robust linking framework. This framework addresses challenges like authentication, which verifies user identity for accessing specific materials, and routing, which directs users to the appropriate content securely. With ISI Links, these processes are efficiently managed within a single centralized system, making it easier for you to obtain the information you require. Through ISI Links, the ISI Web of Knowledge transforms from a collection of isolated resources into a cohesive network, where everything is interconnected. This interconnectedness simplifies your research journey, allowing you to concentrate on finding and utilizing high-quality information for your academic work.

5.1. Comprehensive Linking Capabilities

ISI Links enables various types of connections to enhance the user experience, including:

- **Intra-content links**: These allow users to navigate within a single source, such as moving from a journal article to its "Times Cited" list or the journal's table of contents.

- **Inter-content links**: These facilitate connections between different resources, for instance, linking a journal article to a chemical reaction illustration or connecting a patent record to a related journal article.

- **Publisher full-text links**: Users can access full-text articles, whether they are hosted on the ISI platform or directly by the publisher.

- **Context-sensitive links**: These links are generated by systems like SFX, providing users with relevant connections based on the context of their search.

- **OPAC links**: These links connect users to Online Public Access Catalogs.

- **Links to protein and gene sequence databases**: These facilitate access to important biological data.

- **Pay-per-view links**: These options allow users to purchase access to specific content as needed.

The centralized server of ISI Links consolidates all these linking methods, which helps eliminate the redundancy and costs associated with having separate authentication and access systems for each resource. This integrated approach streamlines the process of verifying user identities (authentication), determining access rights based on subscriptions, and directing users to the correct content.

5.2. Accessing Full-Text: Direct Publisher Connections

As electronic full-text articles became more common online, ISI took advantage of its partnerships with key publishers to create the initial version of its full-text linking system. This first phase used direct electronic feeds from publishers to provide strong and reliable links between ISI's bibliographic records and the full-text articles available on publisher websites. Publishers supplied essential metadata, which included bibliographic details, unique identifiers, and URLs for their articles. Using this information, ISI developed a system that generated specific "keys" to be kept in a links table. This process allowed for the creation of stable hotlinks whenever a match was detected between the publisher's data and ISI's records.

5.3. ISI RoboLinks: Ensuring No Dead Links

To expand access to full-text articles from more publishers and societies, ISI introduced RoboLinks. This innovative system blends direct electronic feeds with dynamic, algorithm-driven linking. These "optimistic" links are created in real-time using metadata, allowing the system to generate URLs on the fly as needed. While optimistic links depend on the accuracy of both metadata and algorithms, RoboLinks adds dependability by pre-verifying links before displaying them. This ensures researchers encounter no dead links, allowing them to navigate directly to entitled full-text resources. RoboLinks accommodates publishers unable to provide direct electronic feeds by allowing ISI to verify links, ensuring flexibility and reducing the burden on publisher sites.

5.4. Customization and Appropriate Copy

ISI Links enables institutions to customize their linking options based on their unique subscription agreements, ensuring that each institution has complete control over the access provided to its users. ISI supports appropriate copy linking to both Internet and intranet holdings, addressing security, maintenance, and technical issues internally or through a mixed-mode approach. Over 250 institutions globally have adopted this capability.

5.5. SFX for Context-Sensitive Linking: Extending Library Resources

Context-sensitive linking software, such as SFX, plays a crucial role in the ISI platform. SFX, built on the OpenURL linking technology created by Herbert van de Sompel and his team, allows institutions to create context-sensitive links between various electronic resources in their libraries. This includes connections between scientific databases, online catalogs, full-text collections, and e-print archives. SFX connects "sources" (systems with link buttons) and "targets" (services to which SFX can link). For example, ISI Web of Science connects directly with the SFX system, passing bibliographic metadata to SFX, which then identifies related electronic items within the institution. This seamless integration creates a fully interlinked environment, linking heterogeneous information resources regardless of format or protocol, thus expanding the reach and utility of institutional digital collections.

Direct publisher feeds and RoboLinks offer an efficient way to access full-text content within the ISI Web of Knowledge platform, ensuring seamless connections to a wide range of articles and publications. Combined with SFX for context-sensitive linking, these tools create a robust, dependable system that enhances the research experience by ensuring seamless access to a wide range of scholarly resources. ISI Links thus represents a significant advancement in the management and utilization of academic and scientific information.

CONCLUSION

The accessibility and manageability of ISI citation data have reached unprecedented levels, solidifying the bibliometric methods pioneered by Eugene Garfield as indispensable tools for analyzing scholarly communication and scientific trends. ISI actively keeps track of trends and advancements in web applications, database technologies, scholarly communication, digital libraries, and the publishing sector. This commitment allows ISI to continually improve its analytical tools for citation data. The recent release of ISI Essential Science Indicators exemplifies ISI's commitment to innovation.

Customer expectations significantly influence ISI's strategic direction. By integrating new technologies with a forward-thinking approach to scholarly communication, ISI has developed the ISI Web of Knowledge, an integrated environment tailored to meet the evolving demands of researchers. This robust platform serves as a single access point to a wide range of research-focused and academic resources, delivering comprehensive multidisciplinary coverage that includes conference proceedings, chemical-related data, journals, web content, and patents. Strategic alliances with information partners enable specialized coverage in fields such as life sciences, agriculture, and engineering, with plans to expand into additional disciplines.

The ISI Web of Knowledge platform also introduces sophisticated search tools for cross-content and web document searches, along with a comprehensive linking gateway that ensures reliable access to appropriate content at all times. This robust linking infrastructure, combined with advanced search technologies, exemplifies the platform's capability to support efficient and effective research processes.

Currently in its early stages, the platform is set to evolve into a comprehensive research portal. By leveraging cutting-edge technology, the ISI Web of Knowledge seeks to fulfill Dr. Garfield's vision that navigating and analyzing citations are essential tools for discovering new ideas and enhancing scientific understanding. The platform's innovative features and strategic partnerships position it as a critical resource for researchers, enhancing the study of scholarly communication and supporting the continuous evolution of scientific inquiry.

Vocabulary (Index of Terms)

Bibliometrics [ˌbɪblioʊˈmɛtrɪks]   The statistical examination of books, articles, and various other types of published materials. It involves the quantification of written communication and is used to evaluate the impact of research and identify trends within the scientific literature.

Citation Index [ˈsaɪteɪʃən ˈɪndɛks]   A database that lists citations between publications, allows users to trace the influence of research articles through the references cited within them. It is essential for locating related works and tracking the impact of research.

Cross-content Searching [krɔs-ˈkɒntɛnt ˈsɜːrtʃɪŋ]

The ability to search across multiple types of content (*e.g.*, journal articles, patents, conference papers) from a single interface, providing a comprehensive view of related information regardless of its format or source.

Hypernavigation [ˈhaɪpərˌnævɪˈgeɪʃən]

The use of hyperlinks to navigate through a digital environment, such as the web. In the context of citation databases, it refers to the ability to move seamlessly between related research articles through linked references.

Multidisciplinary Content [ˌmʌltɪˈdɪsəplɪnɛri ˈkɒntɛnt]

Content that spans multiple academic disciplines, providing a broad range of information that supports interdisciplinary research and discovery.

Probabilistic Search Engine [ˌprɒbəˈbɪlɪstɪk sɜːrtʃ ˈɛndʒɪn]

A search engine that uses probabilistic models to retrieve information, ranking results based on their relevance to the search query. This approach enhances the accuracy and efficiency of searches within large datasets.

RoboLinks [ˈroʊboʊ lɪŋks]

An automated linking system that dynamically generates URLs to direct users to full-text resources. It ensures stable and accurate access to publisher-provided content by verifying links before presenting them to users.

Scientometrics [ˌsaɪəntəˈmɛtrɪks]

The field of study that measures and analyzes scientific literature. It is a subset of bibliometrics focused specifically on the evaluation and quantification of scientific publications.

Web of Science [wɛb ɒv ˈsaɪəns]

A comprehensive, multidisciplinary citation database that provides access to a vast array of research articles and enables users to perform cited reference searching, track citations, and analyze research impact.

Web of Knowledge [wɛb ɒv ˈnɒlɪdʒ]

An integrated research platform that combines various databases, tools, and resources to support comprehensive scholarly research across multiple disciplines. It includes the Web of Science and additional content sources.

COMPREHENSION TASKS

I. Key Terms: Matching

Instructions: Match each term from the left column with the corresponding description from the right column. Write the letter of the correct description next to the term's number. Each description will only match with one term. Good luck!

1. Bibliometrics	**A.** Database for tracking citations between publications
2. Citation Index	**B.** Automated linking system for full-text resources
3. Cross-content Searching	**C.** Navigation through hyperlinks in digital environments
4. Hypernavigation	**D.** Integrated research platform combining multiple databases
5. Multidisciplinary Content	**E.** Statistical analysis of books and publications
6. Probabilistic Search Engine	**F.** Spanning multiple academic disciplines
7. RoboLinks	**G.** Use of probabilistic models for information retrieval
8. Scientometrics	**H.** Analysis and quantification of scientific literature
9. Web of Science	**I.** Multidisciplinary citation database
10. Web of Knowledge	**J.** Comprehensive view across multiple content types

Cell Biology True or False Quiz: Assess whether the following statements are accurate or inaccurate based on your understanding of cell biology concepts. Indicate "True" if the statement is correct and "False" if it is incorrect. Good luck!

1. ----- Eugene Garfield introduced the concept of citation indexes in 1964.

2. ----- The Science Citation Index was initially available in electronic format.

3. ----- The ISI Web of Science was launched in 1997 as a web-based platform.

4. ----- Hypernavigation allows seamless navigation through linked research papers.

5. ----- The ISI Web of Knowledge supports single-disciplinary research.

6. ----- ISI CrossSearch enables searches across articles, papers, and patents.

7. ----- RoboLinks generates direct links to full-text articles without verification.

8. ----- The ISI Web of Knowledge lacks tools for citation-based metrics.

9. ----- Adding BIOSIS Previews enhances biological science resources.

10. ----- ISI Links manages authentication, access, and routing for the platform.

11. ----- SFX technology creates context-sensitive links within institutional resources.

12. ----- Future developments aim to unify access to internal and external resources.

13. ----- Dr. Garfield's vision for citation navigation became real with internet technologies.

14. ----- The ISI Web of Knowledge has no plans to expand beyond life sciences, agriculture, and engineering.

II. Cell Biology Completion Exercise: Fill in the blanks with the appropriate term or phrase to complete each sentence related to cell biology. Write your answers in the space provided. Good luck!

1. The ISI Web of Knowledge was launched in the year ----------.

2. Eugene Garfield is known for pioneering the concept of ---------- indexes.

3. The Science Citation Index (SCI) initially provided comprehensive citation data for the field of ----------.

4. ---------- navigation allows users to seamlessly move through interconnected research papers within the ISI Web of Knowledge platform.

5. The platform's tools for performance metrics include ISI Journal Citation Reports and ---------- Science Indicators.

6. The integration of BIOSIS Previews into the ISI Web of Knowledge enhances resources in the field of ---------- sciences.

7. The ISI Web of Knowledge platform was created to support ---------- research, encompassing multiple disciplines.

8. SFX technology is utilized within the platform to create ---------- -sensitive links.

9. The upcoming developments of the ISI Web of Knowledge aim to unify access to both ---------- and ---------- resources.

10. RoboLinks is a feature that generates direct links to ---------- articles after verification.

III. Multiple Choice: Choose the best answer for each question by circling the corresponding letter.

1. The ISI Web of Knowledge platform was initially launched in which year?

a. 1985

b. 1986

c. 1997

d. 2020

e. 2023

2. The founder of the ISI Web of Knowledge is:

a. Paul Otlet

b. Vannevar Bush

c. Eugene Garfield

d. Tim Berners-Lee

e. John von Neumann

3. Which index is considered the cornerstone of the ISI Web of Knowledge?

a. Social Sciences Reference Index

b. Arts and Humanities Reference Index

c. Science Citation Index

d. New Sources Citation Index

e. Engineering Reference Index

4. The concept of citation indexing was pioneered by:

a. Albert Einstein

b. Eugene Garfield

c. Isaac Newton

d. John Bardeen

e. Rosalind Franklin

5. The integration of BIOSIS Previews into the ISI Web of Knowledge primarily enhances resources in:

a. Chemistry

b. Physics

c. Biological sciences

d. Engineering

e. Mathematics

6. The ISI Web of Knowledge is designed to support:

a. Single-discipline research

b. Multidisciplinary research

c. Only biological research

d. Only physical sciences research

e. Non-academic research

7. One of the key features of the ISI Web of Knowledge is:

a. PDF annotation tools

b. Hypernavigation

c. Direct messaging between researchers

d. Online tutoring

e. Virtual laboratory simulations

8. Essential Science Indicators is a tool within the ISI Web of Knowledge used for:

a. Experiment design

b. Performance metrics

c. Literature review

d. Data storage

e. Code compilation

9. The platform utilizes SFX technology to:

a. Store research data

b. Generate context-sensitive links

c. Provide video tutorials

d. Manage user accounts

e. Automate citation formatting

10. Future developments of the ISI Web of Knowledge aim to:

a. Reduce the number of journals indexed

b. Distinguish access between internal and external resources

c. Unify access to both internal and external resources

d. Focus exclusively on historical research

e. Limit access to government researchers

READING UNDERSTANDING

A. Immunology

The concept of immunity comes from the Latin word "**immunitas**". In ancient Rome, this term referred to the special rights given to senators, allowing them to avoid certain public responsibilities and legal actions while they held office. Historically, immunity has been associated with protection from diseases, particularly infectious diseases. The immune system consists of cells and molecules that work together to protect the body from foreign substances. This protective action is called the immune response.

The main role of the immune system is to guard against harmful microbes, but non-infectious substances can also trigger immune reactions. Interestingly, the same processes that defend the body can sometimes cause tissue damage or disease. Therefore, immunity can be seen as the body's response to any foreign substance, whether it is helpful or harmful, including microbes and large molecules like proteins and polysaccharides. Immunology is the study of these immune reactions and the biological events that happen when the body encounters foreign substances.

The concept of immunity existed long before it was scientifically defined. Ancient civilizations, including those in *Iran*, made significant contributions to early medical practices. For example, the Persian scholar **Avicenna (Ibn Sina)**, often called the "*Father of Early Modern Medicine*" and the "*Prince of Physicians*," made substantial advances in understanding disease and health. In his renowned work, **The Canon of Medicine**, Avicenna discussed various methods for strengthening the body's defenses against diseases, a concept that anticipated modern ideas about immunity. Iranian traditional medicine also employed protective measures against infectious diseases, such as herbal remedies and isolation techniques to prevent the spread of contagions. Decades later, in ancient China, people attempted to protect children from smallpox by having them inhale powders made from the skin of smallpox survivors.

Modern immunology, however, is an experimental science that focuses on understanding immune functions through controlled studies. One of the earliest examples of manipulating the immune system was Edward Jenner's smallpox vaccination. Jenner, an English doctor, noticed that milkmaids who had recovered from cowpox did not get smallpox. Using this knowledge, he injected material from a cowpox sore into a young boy's arm. When exposed to smallpox later, the boy didn't get sick. Jenner's groundbreaking work, published in 1798, led to the wide use of vaccination to protect against infectious diseases. In 1980, the World Health Organization declared smallpox eradicated, proving the power of immunology.

Since the 1960s, our understanding of the immune system has advanced significantly. New techniques, such as growing cells in labs, producing monoclonal antibodies, using recombinant DNA, and creating genetically modified animals like knockout mice, have transformed immunology from a descriptive science into one that explains immune processes in detail using structure and biochemistry.

1. What is the historical origin of the term "immunity"?

A. From Greek philosophers discussing health

B. From Roman senators exempt from certain duties

C. From ancient Chinese medical practices

D. From medieval European doctors

2. How can immunity be more inclusively defined?

A. As a defense against only infectious diseases

B. As a reaction to any foreign substance, infectious or not

C. As protection from physical injuries

D. As a reaction only to proteins and polysaccharides

3. What was Edward Jenner's significant contribution to immunology?

A. Discovering antibiotics

B. Creating the first vaccine against smallpox

C. Developing X-ray crystallography

D. Inventing monoclonal antibodies

4. Why is Jenner's vaccination method still important today?

A. It eradicated measles worldwide.

B. It demonstrated that smallpox could be eradicated through vaccination.

C. It introduced antibiotics to medicine.

D. It discovered the role of DNA in immunity.

5. What major announcement did the World Health Organization make in 1980?

A. The discovery of the first antibody

B. The invention of monoclonal antibody production

C. The eradication of smallpox worldwide

D. The creation of transgenic animals

6. What advancements since the 1960s have transformed our understanding of the immune system?

A. Monoclonal antibody production and recombinant DNA technology

B. Traditional herbal medicine and acupuncture

C. Radiography and chemotherapy

D. Antibiotic development and surgical techniques

7. What did ancient Chinese practices involve to make children resistant to smallpox?

A. Vaccination with cowpox

B. Inhaling powders made from smallpox lesions

C. Drinking herbal concoctions

D. Receiving monoclonal antibodies

8. What is the main role of the immune system?

A. To improve digestive efficiency

B. To defend against infectious microbes

C. To enhance physical strength

D. To regulate body temperature

9. How has immunology evolved as a science?

A. From a descriptive science to one explained in structural and biochemical terms

B. From focusing on herbs to focusing on surgery

C. From studying only animals to studying plants

D. From a theoretical science to one based on folklore

10. What scientific techniques have contributed to the transformation of immunology?

A. Herbal medicine and acupuncture

B. X-ray crystallography and genetically modified animals

C. Radiotherapy and chemotherapy

D. Antibiotic development and nutritional supplements

B. Immune System

The structural arrangement of immune system cells and tissues plays a vital role in producing efficient immune responses. This organization allows the limited number of lymphocytes that target a specific antigen to find and respond to it effectively, no matter where the antigen enters the body.

The adaptive immune response depends on lymphocytes that are specific to antigens, along with accessory cells required for their activation and effector cells that eliminate the antigens. B and T lymphocytes have a wide range of highly specific antigen receptors, playing crucial roles in the specificity and memory of adaptive immune responses. In contrast, Natural Killer (NK) cells are a unique type of lymphocytes that do not have highly diverse antigen receptors and primarily operate within innate immunity. Various surface molecules characterize different subsets of lymphocytes and other leukocytes, following the CD (cluster of differentiation) nomenclature. Both B and T lymphocytes stem from a shared precursor in the bone marrow. While B cell development occurs in the bone marrow, T cell precursors migrate to the thymus for maturation. After maturing, both B and T cells exit these primary lymphoid organs, enter the bloodstream, and infiltrate peripheral lymphoid organs.

Naive B and T cells are mature lymphocytes that have yet to encounter an antigen that would trigger their differentiation. When they do encounter an antigen, they differentiate into effector lymphocytes, which are essential for protective immune responses. Effector B lymphocytes develop into plasma cells that secrete antibodies, while effector T cells include CD4+ helper T cells that produce cytokines and CD8+ cytolytic (or cytotoxic) T lymphocytes (CTLs).

Some descendants of antigen-activated B and T lymphocytes become memory cells that remain inactive for long periods. These memory cells are crucial for mounting quick and enhanced responses during future exposures to the same antigen.

Accessory cells play a crucial role in activating lymphocytes and carrying out effector functions that can be enhanced by humoral or cell-mediated adaptive immune responses. These cells encompass mononuclear phagocytes, dendritic cells, and follicular dendritic cells.

The organs of the immune system are categorized into generative (or primary) organs, where lymphocytes undergo maturation (such as the bone marrow and thymus), and peripheral (or secondary) organs (like lymph nodes and the spleen), where naive lymphocytes become activated by antigens. The bone marrow houses the stem cells responsible for all hematopoietic and lymphoid cells and serves as the maturation site for these cell types, with the exception of T cells.

Lymph nodes serve as critical sites where B and T cells react to antigens gathered by lymph from peripheral tissues. In contrast, the spleen is where lymphocytes engage with blood-borne antigens. Both lymph nodes and the spleen are structured into distinct areas for B cells (follicles) and T cells (parafollicular zones). The T cell areas also contain mature dendritic cells, which are specialized accessory cells responsible for activating naive T cells. In the B cell areas, follicular dendritic cells play a vital role in activating B cells during humoral immune responses to protein antigens. The development of the architecture of secondary lymphoid tissue is dependent on cytokines. The cutaneous immune system consists of specialized groups of accessory cells and lymphocytes that respond to environmental antigens encountered in the skin. Langerhans cells, which are immature dendritic cells located in the epidermis, capture antigens and transport them to draining lymph nodes.

The mucosal immune system consists of specialized groups of lymphocytes and accessory cells organized to enhance interactions with environmental antigens introduced through the respiratory, gastrointestinal, and genitourinary tracts. Examples of tissues within the mucosal immune system include Peyer's patches located in the intestinal wall and the tonsils found in the oropharynx. Lymphocyte recirculation refers to the ongoing movement of lymphocytes throughout the body *via* blood and lymphatic vessels, which is essential for both initiating and carrying out immune responses. Naive T cells typically circulate among various peripheral lymphoid organs, which increases their likelihood of encountering antigens presented by antigen-presenting cells, such as mature dendritic cells. In contrast, memory and effector T cells are more prone to being recruited to peripheral sites of inflammation where microbial antigens are present.

C. Cytokines

Cytokines are signaling proteins released by cells involved in both innate and adaptive immune responses, orchestrating a wide array of functions. These proteins are synthesized in reaction to pathogens and various antigens, and each cytokine can elicit specific responses from immune and inflammatory cells. During the immune response activation phase, cytokines play a crucial role in promoting lymphocyte growth and differentiation. In the later stages of both innate and adaptive immunity, they activate various effector cells tasked with the destruction of pathogens and other antigens. Moreover, cytokines facilitate the development of hematopoietic cells. Clinically, they are significant as therapeutic agents or targets for specific antagonists in a range of immune and inflammatory conditions.

The naming of cytokines often indicates their cellular sources; for example, those secreted by mononuclear phagocytes are known as monokines, while those from lymphocytes are called lymphokines. With the advent of molecular cloning techniques, it has become apparent that the same cytokine can be produced by diverse cells, including lymphocytes, monocytes, and various tissue cells, such as endothelial and epithelial cells. This realization has led to the preference for the broader term "cytokines" to encompass all such mediators. Additionally, since many cytokines are produced by leukocytes (*e.g.*, macrophages or T cells) and influence other leukocytes, they are frequently referred to as interleukins (IL). However, this terminology can be misleading, as many cytokines that specifically act on leukocytes do not fall under the interleukin category, while numerous interleukins may be produced by and act upon non-leukocyte cells. Despite these inconsistencies, the term remains useful; new cytokines are designated with IL numbers (*e.g.*, IL-1, IL-2, *etc.*) to maintain standardized naming conventions.

In contemporary research and clinical practice, cytokines are increasingly employed to either stimulate or suppress inflammatory processes, immune functions, and blood cell formation. In these contexts, they are often categorized as biological response modifiers.

D. Extending Scientific Knowledge

In recent years, advances in immunology have introduced new concepts and technologies that further enhance our understanding of the immune system. One significant development is the use of immune checkpoint inhibitors in cancer therapy. These inhibitors target proteins that inhibit T cells from attacking cancer cells, thereby enhancing the immune system's ability to combat tumors.

Another emerging area is the study of the gut microbiome's role in immune function. Research indicates that the diverse microbial community in the intestines can impact the development and regulation of the immune system, which may in turn influence how the body responds to infections, vaccinations, and autoimmune diseases.

Gene editing technologies, such as CRISPR-Cas9, are also being explored to modify immune cells for therapeutic purposes. For example, CRISPR can be used to engineer T cells, improving their capacity to identify and eliminate cancer cells, thus providing a hopeful strategy for personalized immunotherapy.

Furthermore, advancements in vaccine technology, including mRNA vaccines, have revolutionized the field, as demonstrated by the rapid development of COVID-19 vaccines. These vaccines harness the body's own cells to produce viral proteins that stimulate a robust immune response, providing protection against infections.

The discipline of immunology is constantly advancing, with ongoing investigations focused on decoding the intricate workings of the immune system and creating groundbreaking treatments for various diseases.

1. What is the primary function of the immune system's anatomical organization?

A. To support the growth of immune cells

B. To ensure lymphocytes can effectively locate and respond to antigens

C. To facilitate nutrient absorption

D. To maintain body temperature

2. What are the essential elements of the adaptive immune response?

A. Natural killer cells and macrophages

B. Lymphocytes specific to antigens, supporting cells, and effector cells

C. Red blood cells and platelets

D. Skin and mucosal barriers

3. Which cells are responsible for the specificity and memory of adaptive immune responses?

A. Monocytes and macrophages

B. B and T lymphocytes

C. Eosinophils and basophils

D. Platelets and erythrocytes

4. Where do B and T lymphocytes originate and mature?

A. Spleen and thymus

B. Thymus and bone marrow

C. Spleen and Lymph nodes

D. Skin and mucosal tissues

5. What functions do naive B and T cells serve in the immune response?

A. To immediately attack pathogens

B. To differentiate into effector lymphocytes upon encountering an antigen

C. To secrete antibodies without antigen stimulation

D. To regulate body temperature

6. What do effector B lymphocytes become after activation?

A. Monocytes

B. Plasma cells: Antibody-secreting cells

C. Natural killer cells (NK cells)

D. Dendritic cells

7. What is the function of memory cells in the immune system?

A. To provide immediate immune responses

B. To stay inactive and quickly respond to future encounters with the same antigen

C. To promote inflammation continuously

D. To inhibit immune responses

8. Where are lymphocytes activated by antigens collected from the lymph?

A. Kidney

B. Thyroid

C. Lymph nodes

D. Liver

9. Which specialized cells are present in the T cell regions of lymphoid organs?

A. NK cells

B. Langerhans cells

C. Mature dendritic cells

D. Plasma cells

10. What is the main function of Langerhans cells in the skin?

A. To produce antibodies

B. To trap antigens and transport them to lymph nodes

C. To secrete cytokines

D. To destroy pathogens directly

11. What is the role of cytokines in the immune system?

A. To store genetic information

B. To mediate functions of immune cells and stimulate responses to antigens

C. To transport oxygen

D. To regulate blood pressure

12. What are cytokines produced by lymphocytes commonly called?

A. Monokines

B. Lymphokines

C. Chemokines

D. Interferons

13. How are cytokines increasingly used in clinical settings?

A. To promote tissue regeneration

B. To either promote or reduce inflammation, immune responses, and hematopoiesis

C. To enhance digestive processes

D. To improve cardiovascular health

14. What recent technological advancements have significantly impacted the field of immunology?

A. Herbal medicine and acupuncture

B. CRISPR-Cas9 gene editing and mRNA vaccines

C. Traditional surgery and chemotherapy

D. X-ray imaging and physical therapy

15. What is the significance of immune checkpoint inhibitors in cancer therapy?

A. They enhance the production of red blood cells

B. They inhibit proteins that stop T cells from attacking cancer cells

C. They reduce the risk of infections

D. They promote the healing of wounds

Section 5
Crafting the Scholar's Path - Navigating the Academic Manuscript Journey

Biology for Students, 2025, 281-342

Crafting the Scholar's Path: Navigating the Academic Manuscript Journey

Understanding the Chapter's Focus: A Roadmap

Academic Manuscript Crafting Guide

Overview of SpringerLink

Crafting Tables and Figures

Overview of ISI Web of Knowledge

Writing and Submitting Manuscripts

Overview of Elsevier ScienceDirect

Language Editing Services I

The Art of Referencing

Language Editing Services II

I. SECTION A

1.1. Crafting Effective Tables and Figures: A Twelve-Step Guide

Tables and figures serve a crucial purpose in scientific communication, providing a means to present complex or voluminous data efficiently and to illustrate trends or patterns inherent in the data. They are pivotal components of scholarly manuscripts, often serving as the focal point for readers who delve beyond the abstract.

Prior to commencing the initial draft of your manuscript, it is paramount to meticulously organize the data intended for presentation. This entails not only preparing the tables and figures themselves but also drafting their accompanying titles, and legends, and performing requisite statistical analyses. This preparatory phase ensures that your results are firmly established before proceeding to their interpretation, fostering confidence in the accuracy and integrity of your findings.

Additionally, this stage offers an opportune moment to assess the completeness of your dataset, identifying any gaps or deficiencies that may need addressing.

Furthermore, before embarking on the drafting process, careful consideration should be given to which results directly address the research questions posed and which data may be omitted without compromising the coherence or comprehensiveness of the manuscript. This strategic planning ensures that the manuscript remains focused and aligned with the overarching objectives of the research endeavor.

(1) Determine the appropriate presentation format for your results, considering whether they are best suited for inclusion within the text, or if they would be more effectively communicated through tables or figures.

(2) Exercise discretion in the inclusion of tables and figures, prioritizing only those that convey essential information that cannot be adequately conveyed through textual description alone.

(3) Ensure that the results presented are directly relevant to the research questions posed in the introduction, regardless of whether they align with or contradict the initial hypotheses.

(4) Design each table and figure to be self-explanatory, capable of conveying meaningful information independently of accompanying text.

(5) Sequentially number each figure and table in correspondence with their order of reference in the text, with separate numbering systems for figures and tables.

(6) Arrange tables and figures in a logical sequence that narrates a cohesive storyline, facilitating comprehension and flow for the reader.

(7) Adhere to the formatting guidelines of the target journal, typically placing tables and figures on separate pages following the reference section.

(8) Avoid page breaks within tables or figures, particularly if the journal requires their integration into the main text. Additionally, refrain from wrapping text around tables or figures.

(9) Ensure that all tables and figures referenced in the article text are appropriately cited and discussed within the manuscript.

(10) Obtain necessary permissions from copyright holders, typically publishers, when including previously published tables or figures, and provide appropriate acknowledgment of the source.

(11) Draft table titles and figure legends in the past tense to maintain consistency with the reporting style of scientific manuscripts.

(12) Provide concise descriptions of the content presented in tables and figures within their titles and legends, refraining from offering interpretations or summaries of the results.

1.1.1. Tables

Tables serve as a means to enhance the readability of an article by presenting numeric data in a structured format. They can also synthesize literature, elucidate variables, or convey survey question wordings.

(1) Utilize the table function within Microsoft Word to create tables, avoiding the use of tabs for formatting.

(2) Employ clear column headings and precise table notes to simplify and elucidate the content of the table. The information presented within each column should be comprehensible without reference to the accompanying text.

(3) Consult the journal's guidelines; typically, they require the table title and table to be situated on the same page, with each table presented on a separate page in numerical order.

1.1.2. Figures

Figures are instrumental in conveying primary findings visually, providing impactful representations of data trends or group results. They can effectively communicate processes or simplify the presentation of detailed data.

(1) Ensure each axis of the figure is labeled with units of measurement, clearly identifying the displayed data (*e.g.*, label each line in a graph).

(2) Refer to the journal's specifications; typically, they require figure legends to be listed in numerical order on a separate page, with each figure presented on a separate page in numerical order.

(3) Ensure figures maintain high image quality, minimizing pixelation. Verify the preferred image file type with the journal.

(4) Consider that most journals prefer figures in black and white to minimize publishing costs, reserving color usage for instances where it provides unique information.

(5) Avoid including experimental details in figure legends; these should be detailed in the methods section.

(6) When using photographs of subjects, ensure written, informed consent was obtained prior to capture.

(7) Select the appropriate figure format: line diagrams or scattergrams for numeric independent and dependent variables, bar graphs for numeric dependent variables only, and bar graphs or pie charts for proportions.

1.2. Discussion Section

The discussion section serves to interpret findings, elucidate their implications, and propose avenues for future research, addressing questions posed in the introduction and contextualizing results within existing knowledge.

(1) Organize the discussion from specific to general, aligning findings with literature, theory, and practical implications.

(2) Maintain consistency in terminology, verb tense (present tense), and point of view used in posing questions in the introduction.

(3) Begin by restating the hypothesis and addressing questions posed in the introduction.

(4) Support answers with results, explaining their relevance to expectations and existing literature.

(5) Address all results pertaining to questions, irrespective of statistical significance.

(6) Describe patterns and relationships in major findings, citing relevant literature.

(7) Defend answers, if necessary, by considering opposing viewpoints.

(8) Discuss conflicting explanations of results.

(9) Address unexpected findings transparently.

(10) Identify and discuss limitations without adopting an apologetic tone.

(11) Summarize the principal implications of findings concisely.

(12) Offer no more than two recommendations for further research.

(13) Explain the study's significance in advancing understanding of the problem.

(14) Be comprehensive yet concise, brief, and specific in discussing all pertinent aspects.

1.3. Developing an Effective First Draft of Your Manuscript:

Crafting the initial draft of your manuscript is a pivotal step in the scholarly writing process, requiring meticulous organization and a clear understanding of your target audience and journal requirements.

1. **Consolidate Information:** Gather all necessary data, references, and preliminary drafts of tables and figures to facilitate efficient writing.

2. **Target a Journal:** Determine the journal aligning with your manuscript's focus. Tailor your writing style and content to meet the expectations of the chosen journal's audience.

3. **Commence Writing:** Begin drafting your manuscript, focusing on capturing the main ideas and concepts. Prioritize writing during periods of heightened energy and concentration, minimizing distractions.

4. **Write Rapidly:** Emphasize the swift generation of ideas without concern for grammar or punctuation. Maintain momentum by writing continuously and leaving gaps as needed.

5. **Express in Your Voice:** Infuse your writing with personal style and tone, enhancing clarity and engagement for readers.

6. **Edit Later:** Resist the urge to edit while drafting; instead, focus on generating content. Editing can impede progress and disrupt the flow.

7. **Adhere to Outline:** Follow the outline structure to maintain coherence and relevance. Transition smoothly between topics to ensure logical progression.

8. **Divide into Sections:** Treat each section as a distinct essay, focusing on its specific goals and objectives.

9. **Set Aside First Draft:** Allow your initial draft to rest for at least a day to gain fresh perspective during subsequent revisions.

10. **Revise Thoroughly:** Revisit your draft with a critical eye, refining clarity, coherence, and organization. Ensure each sentence contributes meaningfully to the manuscript's objectives.

11. **Prioritize Clarity and Brevity:** Streamline sentences and paragraphs for enhanced readability, avoiding unnecessary words or complexities.

12. **Maintain Consistency:** Ensure uniformity in writing style and formatting throughout the manuscript, particularly in collaborative works with multiple authors.

1.4. Selecting a Journal

Choosing the appropriate journal is a crucial decision, impacting the dissemination and visibility of your research findings.

1. **Peer Review:** Confirm that the journal employs rigorous peer review processes for publication.

2. **Relevance:** Assess whether the journal regularly publishes papers in your field and aligns with your research focus.

3. **Reputation:** Consider the journal's prestige and impact within your academic community, evaluating factors such as editorial board expertise and citation metrics.

4. **Citation Frequency:** Determine which journals are frequently cited within your discipline, indicating influence and visibility.

5. **Publisher Affiliation:** Society-affiliated journals often possess greater prestige and wider readership compared to independent publications.

6. **Indexing:** Verify if the journal is indexed in major electronic databases, enhancing discoverability and accessibility.

7. **Expertise:** Select journals with relevant expertise to ensure a fair and thorough review process.

8. **Targeted Readership:** Choose journals whose readership aligns with your intended audience and dissemination goals.

9. **Publication Frequency:** Consider the journal's publication frequency and review timeline to align with your desired publication schedule.

10. **Language and Format:** Assess whether the journal publishes in English and evaluate its formatting preferences.

11. **Scope and Focus:** Determine if the journal's scope and research orientation match your manuscript's content and objectives.

12. **Aesthetic Preferences:** Review the journal's formatting style and citation conventions to ensure compatibility with your manuscript's presentation.

Upon selecting a journal, obtain and adhere to the journal's author guidelines meticulously to optimize the manuscript's submission and publication process. With a clear understanding of your research goals, target audience, and chosen journal, you are well-equipped to embark on drafting your manuscript.

II. SECTION B

2.1. Guidelines for Writing and Submitting Manuscripts

Submitting a manuscript to a journal requires careful preparation to adhere to the journal's guidelines. Utilizing a checklist can help ensure that your manuscript meets all requirements for acceptance. Most journals provide guidelines on their website and may update them regularly, so it's essential to review the most recent version. Below is a general checklist to assist in preparing your manuscript, but keep in mind that specific journals may have additional requirements.

2.2. Journal Submission Checklist

i. Cover Letter

❖ Determine if a cover letter is necessary.

❖ Address the appropriate editor based on the manuscript's subject.

❖ Ensure correct addressing.

❖ Review cover letter requirements.

ii. General

❖ Identify the article type being submitted.

❖ Use the specified font type and size.

❖ Adjust line spacing (single or double).

❖ Check format for section headings.

❖ Arrange sections in the correct order.

❖ Verify word length limits.

❖ Include line numbering, if required.

❖ Add page numbers, if required.

❖ Adjust margin size.

❖ Ensure correct nomenclature.

❖ Check for spelling errors.

❖ Determine if results and discussion are separate or combined.

iii. Title Page

❖ Confirm allowed title length.

❖ Determine if a running or short title is needed.

❖ Check for required keywords.

❖ Confirm the need for a list of abbreviations.

❖ Include all authors and their affiliations.

❖ Ensure correct formatting of authors' names and addresses.

❖ Provide corresponding author information.

iv. Abstract

❖ Verify word limit.

❖ Determine if a structured or unstructured abstract is required.

v. References

❖ Ensure correct in-text citation format.

❖ Verify all cited references are included and vice versa.

❖ Check formatting and accuracy of references.

vi. Tables and Figures

❖ Format in-text mentions of tables and figures correctly.

❖ Ensure tables and figures are in the appropriate location.

❖ Use correct fonts and font sizes.

❖ Confirm numbering format for tables and figures.

❖ Check sizing and file formats.

❖ Determine the format for table titles and figure legends.

❖ Ensure all tables and figures are referenced in the text.

❖ Verify guidelines regarding vertical lines in tables.

vii. Other Considerations

❖ Determine if a conflict of interest statement is required.

❖ Check for necessary funding source disclosures.

❖I nclude ethical and patient approval statements for medical manuscripts.

2.3. Promoting Your Publication

Merely publishing an article doesn't guarantee it will be read or cited. With the vast amount of publications daily, scientists may not notice your contribution unless actively seeking it. To ensure your paper reaches relevant academic circles, take proactive steps:

1. **Share Your Work:** Send copies of your paper to academic contacts, including authors cited, researchers in your field, supporters of your research, junior researchers, institute librarians, and relevant interest groups.

2. **Utilize Reprint Services:** Many journals offer reprint services, allowing you to order extra copies for distribution to potential recipients identified in your network.

3. **Tailor for Non-Specialists:** If sharing with policymakers or non-specialists, include a cover letter summarizing your paper in layman's terms, emphasizing its importance.

4. **Optimize Online Discoverability:** Since most papers are found online, use descriptive keywords and titles for easy database searchability.

5. **Create a Downloadable** Repository: Maintain a website where your published work is easily accessible, reducing reprint costs. Inform your contacts to visit your site and download papers.

6. Responding to Reviewers

After manuscript submission, reviewers' comments are provided by the journal's editor. If rejected, assess feedback for possible revisions. If provisionally accepted, plan a strategy for full acceptance:

1. **Review Comments:** Carefully read all comments from reviewers and the editor, taking time to reflect before responding.

2. **Implement Changes:** Incorporate suggestions that improve your manuscript's quality.

3. **Respond Promptly:** If provisionally accepted, draft a detailed response promptly, addressing each comment respectfully.

4. **Be Polite and Clear:** Maintain a polite tone in your response, explaining your viewpoint calmly and clearly.

5. **Address Comments Completely:** Respond to each comment thoroughly, making necessary changes to the manuscript.

6. **Resolve Conflicting Advice:** If reviewers disagree, make informed decisions on which suggestions to follow, explaining your rationale.

7. **Defend Valid Points:** Provide evidence if contesting a reviewer's comment, ensuring clarity and accuracy in your response.

8. **Adapt to Length Requirements:** If asked to shorten the manuscript, be willing to revise accordingly.

9. **Ensure Consistency:** Confirm that changes made align with comments and adhere to journal guidelines.

10. Eleven Reasons Why Manuscripts Are Rejected

Manuscripts may face rejection due to various factors, many of which are avoidable:

1. Poor experimental design.

2. Misalignment with the journal's scope.

3. Subpar English grammar.

4. Lack of clear problem statement.

5. Inadequate description of methods.

6. Overinterpretation of results.

7. Inappropriate statistical analysis.

8. Confusing presentation of data.

9. Unsupported conclusions.

10. Incomplete literature review.

11. Resistance to addressing reviewer suggestions.

III. SECTION C

3.1. Language Editing Services (I)

In the competitive world of scientific publication, ensuring clarity and accuracy in the language is essential for effectively communicating research findings. To meet this demand, several language editing companies offer comprehensive services tailored to the needs of authors seeking to enhance the quality of their manuscripts before submission or publication in peer-reviewed journals.

Here is a brief overview of eight reputable language editing companies:

1. **American Journal Experts**: With a robust team of language experts, American Journal Experts provides comprehensive editing services to researchers worldwide. Their website offers detailed information on their editing process and pricing.

2. **Asia Science Editing:** Specializing in serving the Asian research community, Asia Science Editing offers language editing services tailored to meet the specific needs and requirements of authors in the region.

3. **Diacritech Language Editing Services:** Diacritech Language Editing Services offers a range of editing solutions, including language polishing and manuscript formatting, to assist authors in preparing their manuscripts for publication.

4. **Edanz Editing:** Edanz Editing is known for its professional editing services aimed at improving the clarity and coherence of scientific manuscripts. Authors can access detailed information about their editing process and pricing on their website.

5. **International Science Editing:** With a focus on assisting authors in achieving publication success, International Science Editing offers comprehensive language editing services along with additional support such as journal selection and manuscript formatting.

6. **International Science Editing-China:** Catering specifically to the Chinese research community, International Science Editing-China provides language editing services tailored to meet the linguistic and cultural needs of Chinese authors.

7. **ScienceDocs Editing Services:** ScienceDocs Editing Services offers a wide range of editing solutions, including substantive editing, proofreading, and formatting, to support authors at various stages of the publication process.

8. **SPI Publisher Services:** SPI Publisher Services specializes in providing language editing services to authors across diverse scientific disciplines. Their website offers detailed information on their editing process and pricing options.

By availing themselves of the services offered by these reputable language editing companies, authors can ensure that their manuscripts meet the highest standards of language and clarity, thereby increasing their chances of publication in prestigious peer-reviewed journals.

In the realm of language editing services, two notable agencies stand out:

3.2. Diacritech Language Editing Services (http://www.languageedit.com/)

3.2.1. Overview

Diacritech Language Editing Services provides comprehensive editing solutions to refine the language of your manuscript, ensuring its adherence to international publishing standards. Diacritech Language Editing Services offers comprehensive editing solutions tailored to enhance the quality and readability of scholarly manuscripts. Diacritech is known for its experienced English language and scientific editors, correct pricing, quick turnaround time, and meticulous attention to preserving the author's scientific content and expertise.

3.2.2. Editorial Services Offered:

The editorial team at Diacritech Language Editing Services offers the following services:

(1) Thorough review for spelling and grammar errors

(2) Enhanced and appropriate use of scientific terminology

(3) Editing to eliminate redundancies

(4) Ensuring correct sentence structure

(5) Usage of suitable vocabulary to meet the standards of US/UK English

(6) Rearrangement of sentences and paragraphs as necessary

3.2.3. Why Choose Diacritech Language Editing Services?

Consider Diacritech Language Editing Services for the following reasons:

(1) Experienced English language and scientific editors oversee the copy refinement process.

(2) Their services have been utilized by prestigious societies, university presses, and commercial publishers.

(3) They are recognized for our transparent and fair pricing structure.

(4) Quick turnaround times are guaranteed to meet your publication deadlines.

(5) You will receive comprehensive notifications regarding all changes made using the "Track Changes" feature.

(6) Your manuscript will be fine-tuned while respecting your expertise and the scientific integrity of the content.

3.2.4. Samples:

Explore our sample works to witness the significant improvements made to sentences, ensuring compliance with copy standards. Each sample demonstrates our editors' proficiency in refining sentences for clarity, coherence, and adherence to language conventions.

Below are a few of our sample works on how the same sentence can be re-framed:

Sample 1

ORIGINAL: The traveling disturbance intensity division is basically away from the traveling line the far and near. Different distance reflection traveling activity human disturbance intensity is different, distance is nearer, the disturbance is bigger, distance is farther, the disturbance is smaller.

CORRECTED: The intensity of the traveling disturbance can be divided by the distance of the traveling routes. Different distance reflects different intensities of human disturbance caused by traveling activities; the shorter the distance, the stronger the disturbance, and the longer the distance, the weaker the disturbance.

Sample 2

ORIGINAL: *A. salina* is a species of Branchiopoda crustacean and one of standard experimental organism of marine contamination toxicity testing.

CORRECTED: *A. salina* belongs to Branchiopoda: Crustacea and is one of the standard experi-mental organisms used for toxicity testing to asses levels of contamination in the marine environ-ment.

Sample 3

ORIGINAL: Now already sixteen kinds of antiserum for virus in the fruit trees have made, but the system of the antiserum has the difficulty and the price is expensive.

CORRECTED: Currently, 16 types of antiserum for virus affecting fruit trees have been identified; however, it is difficult to produce antiserums and they are expensive.

Sample 4

ORIGINAL: The annealing procedure at 800°0 for an hour was applied to the sample grown at 630°0.

CORRECTED: Annealing of the sample grown at $630°0$ was carried out at $800°0$ for 1 hour.

Sample 5

ORIGINAL: The single crystal ZnO was used as comparison.

CORRECTED: The single crystal ZnO was used as a reference.

Sample 6

ORIGINAL: Large-scale commercial release of transgenic crops has sparked off intensive debates worldwide regarding their biosafety in the agroecosystem.

CORRECTED: Due to concerns regarding the biosafety of transgenic crops in the agrosystem, the large-scale commercial release of transgenic crops has sparked intensive debates worldwide.

Sample 7

ORIGINAL: According to the visually different hybridization image on array.

CORRECTED: According to the visual interpretation of the different hybridization images on array.

Sample 8

ORIGINAL: Potassium resource is poor in China, and most of potassium fertilizer depends on import (Wang 1996), which consumes a large quantity foreign exchange annually. Lack of soil. K is getting worse and worse, the effective supplement should be found to accommodate expenditure of soil potassium aiming at different soil types and planting system. Wheat straw is abundant organic material, which is ubiquitous and it's organic potassium resource. Straw is returned to soil directly or indirectly, which can constitute a favorable soil environment (Li *et al.* 2002). Most previous works were studied on effect of fertilization and straw to soil on soil physical character, basic nutrient change and improvement of crop yield and quality (Nel *et al.* 1996; Wang *et al.* 1996; Li *et al.* 1998; Lao *et al.* 2002; Turley *et al.* 2003; Liu *et al.* 2005), and experiments were scarcely studied on different forms of soil potassium and potassium contribution to soil.

CORRECTED: China has poor resources of potassium and large quantities of potassium fertilizer are being imported every year (Wang 1996), which consumes foreign exchange in a large amount. As there is a considerable decrease in soil K, an effective solution that can help reduce expenditure on soil K and is suitable for application to different types of soil and planting systems should be found. Wheat straw, which is ubiquitous, is rich in organic material and it is the source of organic potassium. Straw is returned to the soil either directly or indirectly, which can constitute a favorable soil environment (Li *et al.* 2002). Most of the research that was carried out previously has focused on the effect of straw on the fertility of soil, the changes in the physical characteristics of the soil, the changes caused in the levels of basic nutrients, and the improvement in the quality of crop and yield (Nel et al; 1996; Wang *et al.* 1996; Li *et al.* 1998; Lao *et al.* 2002; 'Turley *et al.* 2003; Liu *et al.* 2005). Only a few experiments have focused on different forms of soil potassium and the contribution of potassium to the soil.

3.3. SPI Publisher Services (http://www.prof-editing.com/)

3.3.1. Overview:

In the global landscape of scientific publishing, proficiency in English writing is a significant advantage. Non-native English-speaking authors often face rejection from esteemed journals and books due to language deficiencies, impacting their prospects for career growth and recognition in their respective fields. In response, SPI Publisher Services, a prominent provider of editorial and content production services for STM (scientific, technical, and medical) publishers, introduced Professional Editing Services. This specialized service aims to assist authors in refining their scholarly, scientific, technical, or medical manuscripts by improving the use of English.

Drawing from extensive experience in editorial and content production, SPI Publisher Services has established a network of highly skilled editors in Asia. These editors possess over five years of experience in science and language editing, with some boasting more than a decade of expertise. Many hold master's degrees, are published authors, and maintain affiliations with research institutions in the Philippines and India.

3.3.2. Samples:

Explore samples of editing provided by SPI Publisher Services, showcasing improvements made to enhance clarity and accuracy in manuscripts:

Sample 1

ORIGINAL: The author would like to thank for all of the kindness suggestions and support given by Ms. Ferrer and Mr. Cornes.

CORRECTED: The author would like to thank Ms. Ferrer and Mr. Cornes for their suggestions and support.

Sample 2

ORIGINAL: I have not sent the article to any publication or submission elsewhere of any part of the study; and the manuscript has been read and approved by all the authors and that the criteria for authorship have been met, and there is not any financial or other conflict of interests. **CORRECTED:** This manuscript has been read and approved by all the authors. The criteria for authorship have been met. The authors also do not have any financial interest or any other conflict of interest. Lastly, this paper or any part of this study has not been submitted to any publication yet.

Sample 3

ORIGINAL: Native starch samples have a very different character from amylolitic digestibility point of view.

CORRECTED: Native starch samples have a very different character in terms of amylolytic digestibility.

Sample 4

ORIGINAL: The coming future will show if the managers of the spas would have managed to react to those environmental changes and if the spa enterprises would have learnt to operate on reduced resources conditions.

CORRECTED: The future will show if managers of spas would have managed to react to such environmental changes, and whether spa enterprises would have learned to operate on conditions of reduced resources.

Sample 5

ORIGINAL: These findings underscore the importance of incorporating the spouse both in the management of illness and in research on the illness experience.

CORRECTED: These findings underscore the importance of involving the spouse both in the management of the illness and in understanding the illness experience.

Sample 6

ORIGINAL: In all, 87 patients and their spouses, and three non-partnered patients were interviewed.

CORRECTED: In all, 87 patients and their spouses, and nine non-partnered patients were interviewed.

Sample 7

ORIGINAL: The challenges are presented in Table 2 and they are elaborated on in the discussion that follows.

CORRECTED: These challenges are listed in Table 2 and elaborated in the discussion that follows.

Sample 8

ORIGINAL: LeMone, J. (1993). Human diabetes mellitus. Image J Nurs Sch 28(2), 101-105.

CORRECTED: LeMone, J. (1993). Human diabetes mellitus. Image: Journal of Nursing Scholarship 28(2), 101-105.

Sample 9

ORIGINAL: In December 2004, we tried to contact the remaining 87 patients whose shoulder destiny was unknown.

CORRECTED: In December 2004, we tried to contact the remaining 87 patients whose shoulder dislocation was unknown.

Sample 10

ORIGINAL: And pre and simultaneous treatment with propofol could reduce the up-regulation of TNF-a and ICAM-1 protein expression ($p<0.01$), post-treatment propofol had not such function ($p>0.05$).

CORRECTED: Although pre- and simultaneous treatments with propofol reduced the up-regulation of TNF-a and ICAM-1 expression (c<0.01), post-treatment with propofol did not have this effect (p > 0.05).

IV. SECTION D

4.1. Language Editing Services (II): *San Francisco Edit (https://www.sfedit.net/)*

4.1.1. Who uses San Francisco Edit?

Scientists and academicians worldwide rely on the services offered by San Francisco Edit for manuscript preparation, including journal articles, abstracts, posters, lectures, and grants. Non-native English speakers often find these services particularly valuable when seeking publication in English-language scientific journals.

The success rate of San Francisco Edit is outstanding, with a significant portion of the edited papers being accepted and published in prestigious journals considered among the best globally.

International corporations and small businesses also turn to San Francisco Edit for proofreading, editing, and rewriting a variety of documents such as white papers, annual reports, business plans, marketing materials, owner's manuals, and other documentation.

With over 35 years of combined experience in publishing and editing scientific journal documents, the principals of San Francisco Edit offer expertise in writing, editing, and proofreading grant applications, manuscripts, professional manuals, marketing materials, and various other documentations.

The team at San Francisco Edit collectively possesses over 250 years of experience in writing, editing, and proofreading academic and non-academic manuscripts.

Clients of San Francisco Edit recognize the following key attributes:

1. The provision of the highest quality work available.

2. Access to a substantial team of thoroughly trained editors and writers.

3. Flexibility to handle projects of any size with as many personnel as required.

4.   Possession of diverse scientific and technical expertise.

4.1.2. What type of manuscripts Does "San Francisco Edit" edit besides manuscripts for peer-reviewed journals?

San Francisco Edit edits a wide range of manuscripts, including:

(1) Book Chapters	(8) White Papers
(2) Grant Applications	(9) Presentations
(3) Ph.D. Dissertation	(10) Business Documents
(4) Protocols	(11) Marketing Collateral
(5) Training Manuals	(12) Consumer Documentation
(6) Policy Papers	(13) User Manuals
(7) Resumes	(14) Others

4.1.3. What File Formats Are Supported?

The primary file format supported by San Francisco Edit is Microsoft Word, compatible with both Windows and Macintosh operating systems.

Utilizing Microsoft Word for editing offers several advantages, notably the use of the revisions or track changes feature. This feature enables a clear presentation of deleted and inserted material in distinct formats and colors, providing various options for reviewing and accepting changes. This functionality proves especially beneficial for capturing minor adjustments, such as changes in punctuation, which might otherwise be overlooked.

While Microsoft Word is the preferred format, San Francisco Edit also accepts other file formats. For further information or any inquiries regarding file compatibility, feel free to contact them.

4.1.4. How does San Francisco Edit guarantee quality of service?

While no editing service can offer a guarantee of acceptance for publication, San Francisco Edit boasts a proven track record of consistently delivering high-quality results. Their team comprises seasoned professionals with years of experience, and they have established processes in place to ensure the excellence of their work.

Once they commit to a deadline, they adhere to it faithfully. Additionally, they take full responsibility for any comments made by referees and/or editors regarding the quality of the final product, particularly in cases involving non-native English-speaking scientists publishing in English. If any issues with the English writing of a paper are identified, they claim to offer to re-edit the paper at no extra charge, ensuring that it meets the required standards for publication.

4.1.5 How can we pay for their services?

They offer multiple payment options to accommodate their clients' preferences. You can pay for their services using a credit card, check, or wire transfer. For further details, please refer to their pricing information.

4.1.6. How long will it take for them to respond to your emails?

According to their statement, they usually reply to emails on the same day they're received. You might get a response within five minutes or it could take up to a day, but it won't exceed 24 hours from when your email is received. If it's a weekend or a U.S. national holiday, they'll follow up on the next business day.

4.1.7. What should I understand regarding delivery times?

San Francisco Edit is known for its swift editing and document services. On average, they take 6-8 days to turn around a manuscript, but for urgent requests, it's 3-4 days. They follow a first-come, first-serve policy, meaning jobs are handled in the order they're received, except for rush orders, which are prioritized. For further details, visit their website (https://www.sfedit.net/).

4.1.8. How to plan work schedules:

If you have a mission-critical, time-sensitive request, it's advisable to schedule it with the service provider as far in advance as possible. This ensures that the necessary editing or support capacity will be available when needed, and facilitates the fastest turnaround time possible. While the service provider aims to offer prompt service even without prior notice, they cannot guarantee a delivery time until the manuscript is submitted and the request terms are confirmed. To make a scheduling request, please reach out *via* email.

4.1.9. Can they guarantee your manuscript will be accepted for publication?

Their primary goal and responsibility are to ensure that your manuscript is accurately written in English and effectively presents your research. However, they do not assess the scientific merit of the research itself. The acceptance of your manuscript ultimately depends on the peer review group and/or editor of the publication to which you submit it.

In addition to editing for clarity, grammar, and other language-related aspects, they also provide conceptual comments on the manuscript as necessary. Studies indicate that manuscripts with strong writing and correct English have a higher likelihood of acceptance for publication.

4.1.10.   What is San Francisco Edit's Confidentiality and Privacy Policy?

San Francisco Edit prioritizes the privacy of its customers and the confidentiality of submitted materials. Customer information is strictly protected and is never shared or sold to any third party. Credit card details are collected through a secure website and are not stored on any open system. Once documents are submitted, they are kept securely by San Francisco Edit. Typically, files are retained for three months after the work is completed, after which they are securely destroyed unless otherwise requested by the customer.

4.1.11. Submission Process

San Francisco Edit boasts a team of Ph.D. scientists who possess extensive experience in writing and editing manuscripts intended for publication in peer-reviewed journals. Their editors meticulously edit and refine your manuscript to enhance its presentation of your research. This involves refining the English, style, vocabulary, and overall flow to ensure alignment with the guidelines of your target journal. They are proficient in both US and UK English and provide constructive comments and suggestions for areas of improvement within the manuscript.

4.1.12. Submission Procedures

Manuscripts submitted to San Francisco Edit must be authored in English and formatted for submission to the intended journal or grant review board.

For complex projects, collaboration with the managing editor *via* phone or email discussions may be necessary. However, for most tasks, authors can simply email the work along with instructions, and the managing editor will take it from there.

The most efficient method of sending and receiving documents is electronically, as a Microsoft Word email attachment to editor@sfedit.net. Along with the manuscript, please include the following information:

1. Name and email address of the person submitting the manuscript.

2. Complete business address, including telephone and fax numbers of the person submitting the manuscript.

3. A second email address or an email address of another author to ensure receipt of the edited manuscript.

4. If the manuscript is intended for publication, specify the targeted journal and the society publishing the journal. If no specific journal is targeted, the manuscript will be edited using generally accepted rules.

5. If possible, include references, as they aid in editing the manuscript.

6. If possible, include figures and tables, specifying whether you want them edited or used only as a reference. Edited figures and tables will adhere to journal guidelines and manuscript formatting.

7. If the manuscript has already been submitted to a journal, provide reviewer's comments to help ensure that the manuscript addresses all suggestions. There is no extra cost for reviewing these suggestions. Additionally, if you have a response letter to the reviewers, it can be edited to correspond with changes made in the manuscript.

8. Any other relevant information necessary for editing the manuscript to your satisfaction.

To save costs, references or title pages are not edited unless specifically requested.

Upon receipt of the documents, they will confirm receipt of the manuscript and aim to complete most editing within 6-8 days. The process of working with San Francisco Edit is straightforward but may vary slightly depending on the type of work requested. Typically, you submit a request or document *via* email attachment.

They acknowledge your request, provide a quote and delivery time, and upon your acceptance, proceed with the work.

4.1.13. Submission Steps

1. San Francisco Edit receives your manuscript.

2. We provide you with a quote and estimated delivery time.

3. Upon your acceptance of the quote and delivery time, we proceed with the editing process.

4. After editing, we send the revised manuscript to you for review along with an invoice.

5. Upon your approval of the edits, you confirm acceptance.

6. Payment can be made *via* check, credit card, or bank transfer.

7. Once payment is received, the editing process is considered complete.

V. SECTION E

5.1. The Art of Referencing: A Comprehensive Guide

5.1.1. Introduction

5.1.1.1. The Importance of Referencing

(1) Demonstrating the breadth of your research efforts through references is crucial for academic recognition and scoring.

(2) Drawing upon the insights of established authorities and utilizing data from reputable sources can significantly bolster the persuasiveness of your own arguments.

Citing sources is a fundamental academic practice that ensures transparency and guards against plagiarism, which involves appropriating another person's work.

5.1.2. When to Include References

(1) References should accompany any direct quotations, as well as tables, charts, images, musical compositions, and other borrowed content.

(2) Rewriting or paraphrasing someone else's work also necessitates referencing.

(3) Summarizing another's work requires proper citation as well.

5.1.3. Providing Detailed Information

Thoroughly documenting your sources facilitates transparency and enables others to verify and explore the materials you've used. Omitting references can raise suspicions of plagiarism, potentially undermining the credibility of your work.

5.1.4. Key Considerations for Reference Lists

(1) Accuracy is paramount to avoid confusion or wasted time for markers or tutors attempting to consult your listed sources.

(2) Including all relevant details ensures clarity and aids in identifying the referenced items accurately.

(3) Adhering to a consistent referencing format streamlines the process of locating specific references within your bibliography.

5.1.5. Referencing Systems

Various referencing systems exist, and the choice typically hinges on the guidelines stipulated by your academic institution. The Harvard Referencing System is widely adopted across faculties, but you should consult your student handbook or seek guidance from your tutor or Faculty Learning/Resource Centre for clarity. Additionally, a brief overview of the Vancouver Method is provided for reference.

5.1.6. Essential Components of a Reference

(1) Identify the author(s) or originator(s) of the work, which could include composers, artists, directors, sculptors, or architects depending on the format.

(2) Acknowledge any editors, translators, or arrangers involved in the production process.

(3) Provide the title of the work, along with any relevant subtitles.

(4) Specify the publisher responsible for presenting the work in its physical form.

(5) Include the date of publication or availability.

(6) If known, cite the place of publication.

(7) Detail any physical attributes of the item, such as page numbers or material type (*e.g.*, CD, DVD, poster).

(8) Offer additional information, like web addresses or catalog numbers, to aid in locating the works.

5.2. The Harvard Referencing System

5.2.1. Citing Items in Text

When utilizing the Harvard system, it is essential to include specific details when referencing other works within your assignment.

5.2.1.1. Quoting Directly from Someone Else's Work:

When quoting directly, provide the author(s) followed by the date in round brackets. For instance:

- "As with any investment, working capital exposes the business to risk" (McLaney, 2003).

If there's no author, indicate either that the work is anonymous (Anon) with the date in round brackets, or provide the title followed by the date in round brackets.

When an author has produced more than one work in the same year, use alphabetical letters to distinguish between them:

- *e.g.*, Singh (2004a), Singh (2004b).

5.2.1.2. Referring to or Summarizing:

When referring to or summarizing a work, include both the author(s) and date. For example:

- McLaney (2003) describes how working capital exposes businesses to risk.

5.2.1.3. Citing a Secondary Source:

When citing a secondary source, indicate this in your list of references. Follow the author guidelines mentioned above for citing within the text. For instance:

- Smith, D. (1990)

5.2.1.4. Page Numbers:

Include page numbers (p) or a range of page numbers (pp) after a comma in the bracket with the date. For example:

- Shah (2002, p.33)

- Jones (2000, pp.17-20)

5.3.1. General Rules

5.3.1.1. Authors

- Single Authors: Family name comes first, followed by a comma, space, and then personal name(s) or initial(s). For example, John, Augustus.

- Two Authors: List authors with "&" between them. For example, Mohammed, A. & Khan, J.

- Three Authors: List authors with a comma after the first and "&" after the second. For example, Pryce-Jones, T., Patel, V. & Brown, P.

- More than Three Authors: List only the first named followed by "*et al.*" For example, Hussain, J. *et al.*

5.3.1.2. Authorship Clarifications

Chairpersons of government or other reports, compilers, illustrators (unless their art is a significant part of the work), translators, arrangers, photographers, and writers of prefaces, forewords, or introductions are not considered authors. Instead, use corporate authorship.

5.3.1.3. Corporate Authorship

A corporate author refers to a group taking responsibility for writing a publication, such as a society, professional body, government department, or any other group.

5.3.1.4. Date

Include the publication date. If there are multiple reissues or reprints, give the earliest date. If the date is from a source other than the title page, enclose it in square brackets.

5.3.1.5. Title

The title should be in italics, copied directly from the item. Use [sic] after any mistake in the published title.

5.3.1.6. Edition

Specify the edition if there are different editions of the work.

5.3.1.7. Place

Include the place of publication, with country enclosed in brackets if unclear.

5.3.1.8. Publisher

Provide the publisher's name as it appears on the item.

5.3.1.9. Other Information

Include additional details such as ISBN, physical format, *etc.*

5.3.1.10. Transliteration

Transliterate any non-Roman alphabet characters as necessary, following relevant standards.

5.4.1 Detailed Examples

At the conclusion of your work, it is imperative to include a list of references, commonly referred to as a bibliography, providing thorough documentation of your sources.

5.4.1.1. Books (or Reports)

When citing a book, it is advisable to extract information from both the title page and the *verso* of the title page for accuracy and completeness.

(1) Basic Essentials of a Reference

A comprehensive reference entry typically includes the following elements: author(s) or editor(s), publication date, title, place of publication, and publisher. An exemplary format for such citation is demonstrated below:

Ommati, M. M.	(2026).	Spermatogenesis.	Harlow:	Pearson/Longman
⬆	⬆	⬆	⬆	⬆
Author	*Date of* Publication	*Title*	Place *of* *publication*	*Publisher*

This structure ensures the proper identification and traceability of the referenced material within academic discourse.

(2) Essential Extras Where They Exist

- **Edition:** If the book is not a 1st edition, indicate the edition immediately after the title. For instance:

Ommati, M. M. (2026). Spermatogenesis. 2nd Ed. Harlow: Pearson/Longman.

- **Volume Number:** If utilizing a specific volume from a multi-volume work, specify the volume number after the title. For example:

Heidari, R. (c2003). Mitochondria. Vol. 1: Statics. 5th Ed. Hoboken, N.J.: Wiley.

- **Page Numbers:** If referencing specific pages, include the page numbers after the publisher. For instance:

Ommati, M. M. (2026). Spermatogenesis. 2nd Ed. Harlow: Pearson/Longman, pp. 145-179.

(3) Optional Useful Extras

- **Subtitle:** Include the subtitle if the title alone doesn't provide adequate information about the book's subject. For example:

Ommati, M. M. (2026). Spermatogenesis: Theory and Context. 2nd Ed. Harlow: Pearson/Longman.

- **Series Title and Number:** If the book is part of a series, provide the series title and the book's number within the series. For instance:

Zamiri, M. J., & Ommati, M. M. (2025). The A-M of Feminism. The A to M Guide Series; No. 20. Bentham Science Press.

- **ISBN (International Standard Book Number):** Include the ISBN, a unique identifier for the book, for better identification. Place this information at the end of the reference. For example:

Ommati, M. M., and Heidari R. (2023). Taurine and the Mitochondria: Applications in the pharmacotherapy of human disease. Bentham Science Press. ISBN: 978-981-5124-49-1.

(4) Parts of Books

If citing a chapter within a book, follow the same format as above, but cite the chapter first, followed by the book details. For instance:

Ommati, M. M., Retana-Márquez, R., Najibi, A., and Heidari R. Advances in nanopharmacology: focus on reproduction, endocrinology, developmental alterations, and next-generational effects (Chapter 5). In Ahmadian. E., Cucchiarini, M., Eftekhari, A (Eds.) (2023). Nanopharmacology and nanotoxicology: Clinical implications and methods. Bentham Science Press. Pp. 100-138   (139).   ISBN: 978-981-5079-70-8.   Doi: 10.2174/9789815079692123010008 (*Optional*).

(5) Electronic Books

Treat electronic books similarly to print ones, but include the website address where the work was accessed and the date of access. For example:

Ommati, M. M., and Heidari R. (2023). Taurine and the Mitochondria: Applications in the pharmacotherapy of human disease. Bentham Science press. [Available at: https://benthambooks.com/book/9789815124484/contents/]   [Viewed   on 2024.05.02]. ISBN: 978-981-5124-49-1.

5.4.1.2. Journal Articles

(1) What is a Journal?

A journal is a publication released periodically under the same title, often with a volume and/or part number. Serials can be published annually, quarterly, bimonthly, monthly, weekly, or even daily. Popular serials, such as *Radio Times*, are usually referred to as magazines, whereas more academic publications are known as journals.

When citing an article from a journal, the following information is required: author(s), year, article title, journal title (in italics), volume (if applicable), part or issue (if applicable), date and month (if there is no volume or if there is a volume but no part or issue number), and page numbers. It is essential to highlight that this referencing should be done according to the guidelines of the journal to which you intend to submit your manuscript or other types of writing. Each journal may have specific requirements or preferred formats for citations, and adhering to these standards is crucial for ensuring the acceptance and credibility of your work.

For example The reference should be formatted as follows:

- **Author, Date, Article Title, Journal Title, Volume, Part/Issue, Page Numbers**

For example:

1.Ommati, M. M., Heidari, R., Manthari, R. K., Tikka, S. Ch. J., Niu, R., Sun, Z., Sabouri, S., Zamiri, M. J., Zaker, L., Yuana, Q., Wang, J. Zhang, J., Wang, J. (2019). Paternal exposure to arsenic resulted in oxidative stress, autophagy, and mitochondrial impairments in the HPG axis of pubertal male offspring. *Chemosphere*. Vol. 236: pp. 124325. DOI: 10.1016/j.chemosphere.2019.07.056. SCI: 2.

(2) Citing a Whole Issue of a Journal

Occasionally, you may need to reference an entire issue of a journal, particularly if it is a special issue focused on a specific topic. In such cases, include the following details: journal title (in italics), subtitle relating to the special issue (if any), year, volume (if applicable), part or issue (if applicable, it may say Special Issue on ...), and page numbers of the issue (if they are not numbered from 1 to the end).

For example:

Chemosphere, Environmental Chemistry, 2024, Vol. 358.

(3) Review in a Journal

When citing a review in a journal, you should also provide details of the work being reviewed. The citation should include the author of the review, year, review title, author of the reviewed work, title of the reviewed work, journal title, issue, page number, and DOI (optional).

For example:

Jafari Fakharbad, M., Moshiri, M., Ommati, M. M., Talebi, M., Etemad, L. (2022). A review of basic to clinical studies of the association between hyperammonemia, and methamphetamine. *Naunyn-Schmiedeberg's Archives of Pharmacology.* Vol. 395, pages 921–931. DOI: 10.1007/s00210-022-02248-w.

(4) Electronic Journal Articles

Electronic journal articles should be cited in the same manner as print articles. It may be beneficial to include information about the hosting service (*e.g.*, Swetswise) and the date of access.

For example:

Parry, Sharon & Dunn, Lee. (2022). Benchmarking Sustainable Practices in Urban Development. *Studies in Urban Planning*, Vol. 22, No. 2, pp. 219-234. [Online version *via* Swetswise] [Accessed on 11/07/2023].

VI. SECTION F

6.1. Overview of Elsevier ScienceDirect

ScienceDirect stands as the foremost full-text scientific database globally, providing access to articles and chapters sourced from over 2,500 peer-reviewed journals and over 10,000 books. Its vast content library currently exceeds 9.5 million articles and chapters, expanding steadily with nearly 0.5 million additions annually. To enhance accessibility, Elsevier has digitized much of the pre-1995 journal content, making articles dating back to 1823 accessible from desktops. Since 2003, authors have been contributing supplementary content such as audio and video files, datasets, and other enriching materials, thereby extending research capabilities beyond traditional print formats. The platform boasts advanced search and retrieval functionalities, empowering users to optimize their knowledge discovery processes. New tools facilitate research workflows, enabling early access to content and seamless downloading, sharing, and storage of materials for collaborative endeavors. For further insights into the ScienceDirect platform, readers can consult the ScienceDirect brochure.

6.2. Introduction to the ScienceDirect Info Site

The ScienceDirect Info site (http://www.info.sciencedirect.com) offers librarians and researchers a comprehensive introduction to ScienceDirect's offerings. It presents an overview of the ScienceDirect product and service portfolio, and outlines content coverage, purchasing options, and usage policies. If visitors cannot find answers to their ScienceDirect-related queries on this platform, they are encouraged to reach out for assistance.

6.3. Overview of Elsevier

Elsevier, headquartered in Amsterdam, The Netherlands, stands as the world's largest provider of scientific, technical, and medical information. With over 2,000 journals, books, and secondary databases in its repertoire, Elsevier is a leading player in the field. It is a constituent of the Reed Elsevier plc group, a renowned publisher and information provider operating across scientific, legal, and business-to-business domains. Reed Elsevier is committed to delivering high-quality and adaptable information solutions to professional users, increasingly leveraging the Internet for content delivery.

6.4. What Does ScienceDirect Provide Access to?

(1) Access to over 2,500 peer-reviewed journals in scientific, technical, and medical fields, with a significant portion sourced from Elsevier.

(2) Free access to all abstracts.

(3) More than nine million full-text scientific articles.

(4) An expanding array of Journals, Reference Works, Book Series, Handbooks, and eBooks, along with millions more full-text articles available *via* CrossRef to other publishers' platforms.

6.5. Key Features of ScienceDirect

(1) Comprehensive coverage of science, offering both breadth and depth, supported by a user-friendly search engine with simple and advanced search options.

(2) Personalization features such as favorite journal lists, email alerts, and saved searches.

(3) Free access to abstracts for all Elsevier journal articles, catering to both guests and licensed users.

(4) Provision of Articles in Press, granting early access to peer-reviewed manuscripts prior to formal publication.

(5) Pay-per-view option for single-article purchases by guest users and licensed users seeking access to unsubscribed content.

(6) Free alerting services including tables of contents, searches, and top downloaded articles in specific subject areas.

(7) 24x7 desktop accessibility.

(8) Integrated data and full Cross-Ref compatibility.

(9) Reliable usage reporting compliant with COUNTER standards.

6.6. Elsevier Journals Available on ScienceDirect

ScienceDirect hosts over 2,500 journals, with coverage for many titles extending back to their inaugural volumes. Some journals have been in circulation for over a century. The Info site provides the most up-to-date lists of titles across various collections and packages. Additionally, articles from discontinued or transferred titles are often accessible online for the first time.

6.7. Understanding Missing Issues in ScienceDirect

Although missing issues are rare, with less than 0.1% occurring after 1994, the addition of pre-1995 Backfiles may lead to occasional instances of missing issues on ScienceDirect. Not all Elsevier titles have a complete printed archive, and such collections were not consistently maintained. Efforts are underway to address these gaps, and ScienceDirect apologizes for any inconvenience caused.

Reasons for missing issues may include:

- Issues never received by ScienceDirect, requiring manual resolution by the analyst team.

- Identified issues not yet available online due to technical or production issues, with ongoing efforts to rectify this.

- Challenges in tracing or handling certain issues, resulting in delays in their online availability. These are gradually made accessible on an issue-by-issue basis, though specific timelines cannot be provided.

6.8. Beyond Journals: Additional Full-Text Resources on ScienceDirect

ScienceDirect offers various full-text resources beyond journals:

- Elsevier Reference Works: Authoritative, comprehensive resources integrated into ScienceDirect.

- Elsevier Book Series: Continuously updated full-text resources.

- Elsevier Handbooks: Available across different subject collections, offering valuable research spanning numerous years.

- eBooks: Covering diverse scientific disciplines, including those under esteemed imprints like Pergamon and Academic Press, with over 5,400 eBooks now fully integrated into ScienceDirect's collection.

6.9. Inclusion of Publications from Publishers Other than Elsevier

ScienceDirect hosts publications from other publishers, including the American Psychological Association (APA) and Tsinghua University in China. A complete list of participating publishers can be found on ScienceDirect.

6.10. Components Available Online *via* ScienceDirect

New issues loaded since approximately October 2001 capture all items listed in the printed table of contents, retrievable *via* ScienceDirect. Depending on the article type and production process, full-text HTML and supplementary data, such as multimedia files, are available. Articles in Press, early versions of peer-reviewed manuscripts, are also accessible.

6.11. Citation Alerts

Citation Alerts notify users *via* email when an article they select is cited by new articles added to ScienceDirect. This feature enables authors to track references to their work and create alerts easily.

6.12. Volume/Issue Alerts

Volume/Issue Alerts inform users *via* email when new volumes of books or issues of journals become available on ScienceDirect. Users can create and manage alerts conveniently through the platform.

VII. Section G

7.1. ISI Web of Knowledge Overview

ISI Web of KnowledgeSM stands as the premier research platform, providing researchers with the ability to swiftly find, analyze, and share information across various disciplines, including the sciences, social sciences, arts, and humanities. The platform ensures integrated access to high-quality literature through a unified interface that connects a diverse array of content and search terms, creating a seamless search experience with a common vocabulary.

With over 20 million researchers in 90 countries relying on the ISI Web of KnowledgeSM for research, planning, and budgeting decisions, the platform's impact is substantial. According to Ms. Elenera Almeida, Access Coordinator of the CAPES Consortium in Brazil, "ISI Web of KnowledgeSM has greatly accelerated our researchers' ability to search for the information they need in their respective fields without useless data returns that only clutter the process. When you save time, you inevitably save money."

7.2. Frequently Asked Questions

7.1.1. What's the Difference Between the Web of Science and ISI Web of Knowledge?

ISI Web of Knowledge is a comprehensive research platform that integrates various types of content, including journal articles, patents, websites, conference proceedings, and Open Access material, accessible through a unified interface. Web of Science®, a component within ISI Web of Knowledge, specifically provides access to journal articles in the sciences, social sciences, and arts and humanities, with over 100 years of valuable research fully indexed and cross-searchable.

7.1.2. Is the Journal I'm Reading or Publishing Included in the ISI Web of Knowledge?

You can browse the Master Journal List at http://scientific.thomson.com/mjl/ to see if your journal is indexed in the ISI Web of Knowledge.

7.1.3. What Is the All Databases Tab in the ISI Web of Knowledge?

The All Databases tab allows simultaneous searches across all content your institution subscribes to within ISI Web of Knowledge, ensuring complete results in one interface without missing any articles, thus providing speed without compromising quality or accuracy.

7.1.4. How Do I Find Specific Articles by Name, Author, or Subject?

In the ISI Web of Knowledge, under the All Databases tab, you can search by Topic, Title, Author, Publication Name, Year Published, and Address. In Web of Science, additional search options include Group Author, Conference, Language, and Document Type. The "Author Finder" function helps locate specific authors within individual databases.

7.1.5. How Do I Find Journals by Specific Subject or Field?

You can use the Master Journal List to find journals categorized by database subjects. Within the ISI Web of Knowledge, registered users can utilize the "My Journal List" feature to search for journals by name, alphabetically, or by subject, and set up Table of Contents email alerts.

7.1.6. What Is the Analyze Tool?

The Analyze tool helps identify trends and patterns in search results, allowing users to discover top authors, institutions, and journals in their area of interest, and observe broad trends in specific topics.

7.1.7. What Is Cited Reference Searching?

Cited reference searching allows users to use a reference's citations to find more articles on the same topic. Users can trace an item's influences backward in time or track forward citations to discover new developments.

7.1.8. How Can I Use the ISI Web of Knowledge to Manage and Format My References and Write Papers?

EndNote® Web, fully integrated into ISI Web of Knowledge, is a bibliographic management tool that enables users to send references, edit them, link to other features, and share folders without leaving the search session.

7.1.9. What Is an Alert?

Users can save search histories as email alerts, manage them *via* their ISI Web of Knowledge homepage, and link directly to full bibliographic records from the HTML emails received.

7.1.10. What Is a Citation Alert?

A citation alert notifies users by email whenever a chosen record is cited by a new record added to the database. Users can create citation alerts by performing a search, viewing a Full Record, and clicking the "Create Citation Alert" button.

By integrating advanced technologies and maintaining a forward-thinking approach to scholarly communication, the ISI Web of Knowledge continues to enhance the research process, supporting researchers worldwide in their pursuit of knowledge and innovation.

ISI Web of Science

Web of Science® offers researchers, administrators, faculty, and students rapid and robust access to the world's leading citation databases. With authoritative, multidisciplinary content covering over 10,000 high-impact journals worldwide, including Open Access journals and over 110,000 conference proceedings, it provides comprehensive and retrospective coverage in the sciences, social sciences, arts, and humanities dating back to 1900.

According to Sul H. Lee, Dean of University Libraries at the University of Oklahoma, "Web of Science makes it possible to conduct cross-disciplinary research and 'drill down' into very specialized subfields within disciplines. The ability to navigate forward or backward within a field of literature, identifying citation patterns and core publications—long a key feature of citation indices—is incredibly easy with Web of Science."

7.2. Key Benefits of Web of Science

7.2.1. High-Impact Articles and Conference Proceedings

- Quickly find articles and conference proceedings of high impact.

- Uncover relevant results across related fields.

- Identify emerging trends, aiding in successful research and grant acquisition.

- Discover potential collaborators with significant citation records.

- Integrate searching, writing, and bibliography creation into one streamlined process.

7.2.2. Why Choose Web of Science?

- **Comprehensive and Relevant Coverage**

Each journal included has passed an objective evaluation process ensuring accuracy, relevance, and timeliness.

- **Cited Reference Searching**

Track prior research, monitor current developments, see who cites your work, measure colleagues' work influence, and follow today's hottest ideas by navigating forward and backward through literature.

- **Easy Author Identification**

Locate articles by the same authors easily, overcoming challenges with similar or identical author names.

- **Insightful Analysis Options**

Discover hidden trends and patterns, gain insights into emerging research fields, and identify leading researchers, institutions, and journals using the Analyze Tool. Citation Reports and Citation Maps visualize citation connections and relationships.

- **Wide-Ranging Proceedings Content**

Track the impact of proceedings papers, conferences, or conference series, particularly valuable in fields like computer science, engineering, and physical sciences.

- **Over 100 Years of Backfile Data**

Conduct deep, comprehensive searches and track trends over a century of vital data.

7.2.3. *What's Included?*

- **Cover-to-Cover Indexing**

Access significant items from journals, including original research articles, reviews, editorials, chronologies, abstracts, and more.

- **A Full Range of Disciplines**

- Information spans disciplines such as agriculture, biological sciences, engineering, medical and life sciences, physical and chemical sciences, anthropology, law, library sciences, architecture, dance, music, film, and theater.

- Access six comprehensive citation databases:

1. Science Citation Index Expanded®: Over 7,100 major journals across 150 disciplines, back to 1900.

2. Social Sciences Citation Index®: Over 2,474 journals across 50 social science disciplines, plus 3,500 leading scientific and technical journals, back to 1956.

3. Arts & Humanities Citation Index®: Over 1,395 arts and humanities journals, plus selected items from over 6,000 scientific and social sciences journals.

4. Conference Proceedings Citation Index: Over 110,000 journals and book-based proceedings in Science and Social Science and Humanities editions, across 256 disciplines.

5. Index Chemicus®: Over 2.6 million compounds, back to 1993.

6. Current Chemical Reactions®: Over one million reactions, back to 1986, plus INPI archives from 1840 to 1985.

- **Integrated Bibliography Management**

EndNote Web® integration for easy access and organization of references online. Send, share, and store references between search sessions.

- **Related Records®**

Enhance cited reference searching across disciplines by finding articles with common references.

- **Times Cited**

Discover a paper's influence by linking to all citing papers.

- **Full-Text Links**

Direct access to full-text articles from publishers.

- **Alerting and RSS Feeds**

Save searches as email alerts or RSS feeds to stay updated with the latest information.

By leveraging the capabilities of Web of Science®, researchers can efficiently navigate the vast landscape of scholarly literature, identify key trends and influential works, and manage their research effectively with integrated tools and comprehensive data coverage.

7.3. Journal Citation Reports (JCR)

Journal Citation Reports® (JCR) offers a systematic, objective means to critically evaluate the world's leading journals, with quantifiable, statistical information based on citation data. By compiling articles' cited references, JCR Web helps measure research influence and impact at the journal and category levels, showing the relationship between citing and cited journals. Available in Science and Social Sciences editions, it remains an essential tool for academic and professional communities.

According to Peter Jacso from Peter's Digital Reference Shelf, "JCR is still the only usable tool to rank thousands of scholarly and professional journals within their discipline or subdiscipline. For educated decisions about selecting and deselecting journals in college libraries, and gauging the prestige and influence of journals, it is a very good tool."

7.3.1. Key Features and Benefits

7.3.1.1. Support and Evaluation for Librarians

Evaluate and document the value of library research investments.

7.3.1.2. Influence Measurement for Publishers

Determine a journal's market influence, review editorial policies, monitor competitors, and identify new opportunities.

7.3.1.3. Publishing Guidance for Authors and Editors

Identify the most appropriate and influential journals for publication.

7.3.1.4. Research Discovery for Researchers

Find current reading lists and track bibliometric and citation trends and patterns.

Why Choose Journal Citation Reports?

Clear Data Sorting

- Sort journal data by Impact Factor, Immediacy Index, Total Cites, Total Articles, Cited Half-Life, or Journal Title.

- Sort subject category data by Total Cites, Median Impact Factor, Aggregate Impact Factor, Aggregate Immediacy Index, Aggregated Cited Half-Life, Number of Journals in Category, and Number of Articles in Category.

Five-Year Impact Factor Trend Graph

View a journal's impact over five years.

Citation Influence and Prestige

Understand citation influence and prestige with Eigenfactor™ Metrics, a five-year metric considering scholarly literature as a network of journal-to-journal relationships.

Impact Factor Visualization

Visualize impact factor by journal category with impact factor boxplots.

Rank Multiple Categories

Rank journals across multiple categories.

Self-Citation Analysis

See how journal self-citations affect the impact factor.

Integration with ISI Web of Knowledge

Link from Web of Science to JCR Web, from JCR journal records to ulrichsweb.com, and to and from your library's OPAC.

Included Features

In-Depth Journal Coverage

- Covers more than 8,000 of the world's most highly cited, peer-reviewed journals from 3,300 publishers across 227 disciplines in 66 countries.

- Available in two editions: Sciences Edition covers over 6,500 journals; Social Sciences Edition covers over 1,900 journals.

Historical Data

Cited and citing journal statistics from 1997 onward.

7.3.1.5. Understanding Impact Factor

Journal Impact Factor

The average number of times articles from a journal published in the past two years have been cited in the JCR year. It is calculated by dividing the number of citations in the JCR year by the total number of articles published in the two previous years.

5-Year Journal Impact Factor

Similar to the Journal Impact Factor but considers a five-year period.

Aggregate Impact Factor

Calculated for a subject category, it accounts for citations to all journals in the category and the number of articles from all journals in the category. This mitigates the advantages of larger or older journals, making it a valuable evaluation tool.

Impact Factor Trend Graph

Shows the Impact Factor for a five-year period.

Impact Factor Box Plot

Depicts the distribution of Impact Factors for all journals in a category, with interquartile ranges and outliers shown.

By leveraging the capabilities of JCR, researchers, librarians, publishers, and authors can gain crucial insights into journal performance, influence, and trends, aiding in informed decision-making and strategic planning in scholarly communication and research.

7.4. Swine Influenza: Lessons from History Identified by the Web of Science®

The Web of Science, encompassing over 11,000 scholarly journals and 130,000 academic conferences worldwide, serves as a paramount citation database. Leveraging its unique citation coverage, this article aims to uncover crucial historical research relevant to investigating the current Swine Influenza outbreak.

7.4.1. Historic Trends

Utilizing a Boolean search in the Web of Science for "(influenza or flu) and epidem*," 1,477 unique articles were identified. A significant surge in publications occurred shortly after the A/H1N1 influenza (Spanish Flu) outbreak in 1918-1919. While the 2009 Swine Influenza outbreak may not match the virulence of the 1918 epidemic, it shares the A/H1N1 strain. Similar peaks in scholarly articles followed outbreaks of the Asian Flu (A/H2N2 strain) in 1957 and the Hong Kong flu (A/H3N3 strain) in 1968. Recent research activity may also stem from outbreaks of the H5N1 avian influenza strain, notably in 2003 and 2005.

7.4.2. Highly Cited Articles

The number of citations to an article serves as a measure of its influence and significance. Several highly cited articles from the Web of Science related to influenza epidemics between 1918 and 1980 are highlighted, demonstrating their enduring value to current research.

- J. Houswort, A.D. Langmuir, "Excess mortality from epidemic influenza," American Journal of Epidemiology, 1974.

- T.C. Eickhoff, LL. Sherman, IL, R.E. Serfling, "Observations on excess mortality associated with epidemic influenza," JAMA, 1961.

- F.M. Davenport, A.V Hennessy, T. Francis, "Epidemiologic and immunologic significance of age distribution of antibody to antigenic variants of influenza virus," Journal of Experimental Medicine, 1953.

- F.L. Horsfall, "Neutralization of epidemic influenza virus," Journal of Experimental Medicine, 1939.

7.4.3. Conclusion

While influenza epidemics are infrequent, historical research remains invaluable to current investigations. However, accessing and identifying relevant articles from older literature can be challenging. The Web of Science, with its comprehensive retrospective coverage and advanced search capabilities based on citations, proves instrumental in unearthing this invaluable yet often overlooked information.

VIII. SECTION G

8.1. Springer Link

SpringerLink serves as an integrated full-text database housing a vast collection of journals and books published by Springer. Presently, SpringerLink boasts over 1,250 fully peer-reviewed journals and an extensive selection of more than 10,000 books accessible online.

For comprehensive insights into our journals, books, and software offerings, visitors can explore detailed information available on our corporate website, springer.com.

To streamline navigation and address commonly asked questions, visitors can select from the following topics to access relevant FAQs:

- Journal Subscriptions

- Book Purchases and Access

- Software Licensing and Downloads

- Platform Features and Functionality

- Account Management and Support

By offering dedicated FAQs tailored to specific topics, SpringerLink aims to enhance user experience and provide efficient assistance to our valued users.

8.1.1. Navigating SpringerLink

SpringerLink provides users with complimentary access to a range of features, including search functionalities, tables of content, abstracts, and alerting services. Full-text access to articles published in your preferred journals is typically available through institutional subscriptions or consortium memberships. Additionally, personal subscribers enjoy online access as part of their subscription benefits.

For further details on new journal releases, ebooks, licensing opportunities, and more, visitors are encouraged to explore the resources available at springer.com/librarians. This dedicated section offers comprehensive information tailored to the needs of librarians and institutions.

8.1.2. Accessing Content on SpringerLink

8.1.1.1. Accessing Full-Text Articles

To access full-text articles within a journal on SpringerLink, you or your institution must be a registered subscriber to that particular journal. Similarly, to access full-text articles within a book series on SpringerLink, you or your institution should have either a standing order to the series or an electronic-only subscription.

For comprehensive details regarding product offerings and licensing options, please visit springeronline.com/librarian. This resource provides in-depth information tailored to librarians and institutions seeking access to Springer content.

8.1.1.2. Advantages of Registering with SpringerLink

Registering with SpringerLink offers numerous benefits for individual users. By registering, you gain access to special personalization features tailored to enhance your user experience. These features include new issue and keyword alerting, Favorites linking for easy access to preferred content, and the ability to access free samples.

Moreover, registration enables you to purchase articles directly on SpringerLink. All purchased articles are conveniently stored in your account's "Order History," ensuring easy retrieval whenever needed. Registering with SpringerLink empowers users to maximize their interaction with the platform and optimize their access to scholarly content.

8.1.1.3. Registration Process for New Customers on SpringerLink

Registering with SpringerLink as a new customer is a straightforward and free process. Begin by accessing the SpringerLink homepage and locate the "Register" button positioned at the bottom-right corner beneath the "Login" box. Click on the "Register" button to proceed.

Upon clicking, you will be directed to a registration page where you can choose to register as an "Individual." Next, fill in the required contact information and create a username and password for future access to the site. Once completed, you'll be prompted to review and accept the terms and conditions for using Springer's electronic content.

After submitting your registration details, an email containing your login credentials will be automatically sent to the provided email address. This email will confirm your registration and provide an eight-digit identification number (Metapress ID) for future reference.

Returning to the SpringerLink homepage, you will now be recognized as a registered user, granting you access to the platform's special features. If you have a valid subscription for a Springer electronic journal or book, it's essential to notify the SpringerLink helpdesk to activate your electronic access. Provide your new Metapress ID, subscription title, and order number if available to expedite the process.

8.1.1.4. Registering with SpringerLink as an Institution

The process of registering with SpringerLink as a new institution is both free and straightforward. Begin by accessing the SpringerLink homepage and locate the "Register" button positioned at the bottom-right corner beneath the "Login" box. Click on the "Register" button to proceed.

Upon clicking, you will be directed to a registration page where you can select to register as an "Institution." Fill in the required contact information for your institution, including creating an institution code, typically using alpha letters representing your institution as an acronym.

The username will be automatically generated based on the institution code and tagged as "admin." This username is for the administrator of the SpringerLink account, who will serve as the main contact for all electronic subscriptions.

Next, you'll have the opportunity to set up authentication methods, including username and password authentication for the administrator side and IP authentication for multiple users.

After completing the electronic registration, you will receive an eight-digit MetaPress ID, which you should keep for reference as it will be required for the contract.

Clicking "Continue" will lead you to the contract information page for institutional customers. Here, you'll find the link to open and print the Springer electronic publications contract. This contract must be completed, signed, and either mailed or faxed to Springer within two weeks of registration.

Once Springer receives and processes the contract (usually within 1-3 business days), you will receive a welcome email from the SpringerLink helpdesk.

Upon electronic registration completion, you can return to the SpringerLink homepage by clicking "Continue." However, please note that access to full-text articles for your institution's titles will only be granted after faxing the contract and receiving the welcome email from the SpringerLink helpdesk.

8.1.1.5. Setting Up Institution Access via Athens

If your institution is part of a Springer consortium, patrons can utilize Athens to access SpringerLink. Patrons simply need to click on the "Login *via* Athens" link located on the right side of the SpringerLink homepage and enter their Athens username and password.

To enable access *via* Athens for your institution, reach out to the SpringerLink Helpdesk at springerlink@springer.com. Provide them with your Athens organization ID and MetaPress ID.

Upon receiving your request, the SpringerLink Helpdesk will coordinate with Athens to activate access for your institutional patrons on SpringerLink. Once Athens completes the setup, the SpringerLink helpdesk will inform you, confirming the activation of authentication.

8.1.1.6. Assistance with Access and Registration Inquiries

For any questions or concerns regarding access and registration, feel free to reach out to the SpringerLink helpdesk:

- For customers outside the Americas, please contact: springerlink@springer.com

- For customers located within the Americas, please contact: springerlink-ny@springer.com

Our dedicated helpdesk teams are here to assist you promptly and efficiently with any inquiries you may have.

8.2. Author Information

8.2.1. Locating Your Published Article on SpringerLink

As an author, discovering your article on SpringerLink is a straightforward process. Upon the online publication of your article, you will promptly receive a Springer Author Alert email. This email, sent on the day of your article's final version publication, contains a direct link to the abstract page of your article, ensuring immediate access.

Additionally, you can swiftly locate your article on SpringerLink using the general search feature located at the top of each page. We recommend entering either the article title or your name as the author in the search bar, and selecting "Articles" in the "Return" field for optimal results.

For a more targeted search, you may utilize the "Advanced Search" option available at the top of every page on SpringerLink. This feature allows you to search "Articles by Citation," providing the flexibility to include the publication name in your search criteria, enhancing the accuracy of your search results.

8.2.2. Linking to Your Online Article as an Author

To effectively link to your article once it's available online, it's advisable to utilize the Digital Object Identifier (DOI). Follow these steps to ensure accurate linking:

1. Access the abstract page of your article on SpringerLink.

2. On the article abstract page, locate and click on "Linking Options" in blue, positioned on the right side of the page.

3. Under the "OpenURL linking" section, you'll find the link labeled below "To link with the Digital Object Identifier (DOI), use." This link is specifically designated for linking to your article.

4. Copy this DOI link and paste it into the URL address bar in your browser window. You can also share this link with colleagues to enable them to view your article effortlessly.

8.2.3. Accessing Your Article in Electronic Format as an Author

To obtain your article in electronic format, follow these steps:

Affiliation with a Research Institution: Typically, you should have access to the final full-text version of your article through your affiliation with a research institution. Check with your institution's library or database access to obtain the electronic version.

Individual Registration on SpringerLink: If you do not have access through an institution, register as an individual user on SpringerLink.

Locate Your Article: Find your article on SpringerLink by searching for the title. Click on the article title to access its abstract page.

Purchase Option: On the article abstract page, select the "Purchase Item Now" option to buy the article. The transaction is secure.

Choose Format: Depending on availability, you can purchase the article in PDF, HTML, or both formats.

Access Your Purchased Article: After completing the transaction, access your article by clicking on "Order History" on the blue banner toolbar. Your purchased articles will be stored here and accessible every time you log in to SpringerLink.

Note: Online First articles are initially available only in HTML format and later provided in PDF format upon publication in a paginated print issue.

8.2.4. *Exploring Journal Information*

To find information about a journal, follow these steps:

1. Visit the Journal Homepage: Go to the homepage of the journal you are interested in.

2. Publisher Information Links: Look for blue links on the right side of the page. These links lead to our publisher information site, Springer Online.

3. Access General Information: Click on the provided links to access general information about the journal, including details about the editorial board and manuscript submission instructions.

4. Visit Springer's Author Page: Alternatively, you can directly visit Springer's Author page, where you'll find comprehensive information about journals, editorial boards, and submission guidelines.

By following these steps, you can easily access the information you need about the journal of your interest.

8.3. Troubleshooting Error Messages

Encountering an error message while trying to open an article can be frustrating, but we're here to help. Follow these steps:

1. Notify the Helpdesk: Immediately inform the SpringerLink helpdesk about the error message you received. Be sure to provide specific details about the error for better assistance.

2. Specify the Error: Describe the type of error you encountered when attempting to access the article. The more detailed your description, the easier it will be for our staff to address the issue.

3. Assistance from SpringerLink: Our dedicated staff will take action promptly. They will either correct the link causing the error or reach out to the typesetters to ensure you receive a working PDF.

By reporting the error and providing detailed information, we can quickly resolve the issue and ensure you have seamless access to the desired article.

8.3.1. Assistance with Error Messages

If you encounter an error message while accessing an article, don't hesitate to reach out for support. Here's how you can get assistance:

1. Email Support: Send an email to the SpringerLink helpdesk with details including the article title, your MetaPress ID, and if possible, a screenshot of the error message page. Use the appropriate email address based on your location:

- For customers outside the Americas: springerlink@springer.com

- For customers inside the Americas: springerlink-ny@springer.com

2. Feedback Form: Alternatively, you can use our Feedback Form, accessible in the blue header information at the top of every SpringerLink page. Provide as much detail as possible to help us address the issue efficiently.

Our team is dedicated to resolving any issues you encounter promptly, ensuring you have a seamless experience accessing our content.

8.4. Accessing Free Sample Issues

8.4.1. How Do I Request and View Free "Sample Issues

If you're keen on exploring specific issues offered by SpringerLink without any cost, our "Sample Issues" feature allows just that. Here's how to request and view these free samples:

1. **Log In or Register**: Ensure you're logged into your SpringerLink account. If you're not registered yet, it's a quick process to sign up and access this special feature.

2. **Requesting a Sample**: Navigate to the journal homepage where volumes and issues are listed. Look for the "request a sample" link **next** to the desired issue.

3. **Confirmation**: Clicking on this link will lead you to a confirmation page. Simply click on the provided link to confirm your **request**.

4. **Accessing the Sample**: Once confirmed, return to the journal homepage. You'll now notice the free "Sample Issue" **highlighted** prominently. Click on it to gain free online access.

5. **Exploring Contents**: Dive into the Table of Contents page of the sample issue. Look out for the eyeglass icon, indicating full **text** access. Click on any title to view the article abstract. From there, click on "Open Full Text" to access articles in PDF or HTML format.

This feature allows researchers and authors alike to preview content and formatting preferences, aiding in informed decisions about subscriptions and research directions.

8.5. Librarian Information

8.5.1. Linking to Journals and Book Series on SpringerLink

Linking to content on SpringerLink is made easy through systematic OpenURL linking. Here are three options for obtaining OpenURL links for SpringerLink content:

Downloading Complete Lists: From the SpringerLink homepage, under the "Browse and Explore" section, click on "Please click here to download a list of journal URLs." This will direct you to the "Linking Exports" page where you can generate lists of URLs in Excel spreadsheet or Tab Delimited text formats. You can also create subject-specific lists known as SpringerLink "Online Libraries" from this page.

Specific Linking at Different Levels: SpringerLink offers specific linking at three levels: journal, issue, or article. Simply navigate to the respective homepage or abstract page and click on "Linking Options" on the right of the page. Choose between MetaPress Direct Linking or Open URL Linking.

Updating Your Library's Catalog: To update your library's catalog listing of URLs for subscribed journals, log in as the administrator of your institution. From the SpringerLink homepage, click on "Please click here to download a list of journal URLs" under the "Browse and Explore" section. On the "Linking Exports" page, select the radio tab next to "My Subscribed Journals" to generate a specific list of your library's subscriptions.

Remember, copying links directly from the browser won't work as they're not permanent links and will redirect to the SpringerLink homepage in the future. These options streamline the process of linking to SpringerLink content for librarians and institutions.

8.5.2. Finding More Information about Springer for Librarians

For librarians seeking detailed product and service information, we recommend visiting our dedicated librarian page on our corporate website, Springer Online. This resource provides comprehensive insights tailored specifically for librarians, offering valuable information to support your institution's needs.

8.5.3. Springer's Inter-Library Loan (ILL) Policy

Springer permits authorized users to produce print copies of articles for inter-library loan (ILL) requests for academic, research, or other non-commercial libraries. However, digital transmission of articles is not permitted under this policy. Authorized users are also allowed to print and store reasonable numbers of individual articles for educational purposes and offline review.

8.5.4. Using Electronic Versions of Springer Journals and Articles for ILL Requests

Electronic versions of Springer journals and articles can be utilized to print copies for ILL requests. These printed copies can then be sent *via* post, fax, or fax-based services (*e.g.*, Ariel or Prospero) to fulfill requests from academic, research, or other non-commercial libraries. However, sending digital transmissions of articles *via* email for ILL purposes is not permitted.

8.5.5. Contact for Online Usage Reports

For inquiries regarding online usage reports for subscribed titles, please reach out to your Springer licensing manager. You can find their contact details on our librarian page at springeronline.com.

8.5.6. Accessing Full Text of Articles from 1978 and Earlier

If you are seeking access to an article published in 1978 or earlier and can only view the abstract, you may benefit from the SpringerLink Historical Archive Project. This initiative aims to digitize content from 1843 to 1996, previously available only in print. SpringerLink now provides electronic access to this archival content, which is indicated by a distinct journal homepage and labeled as Historical Archive.

While abstracts for all Historical Archive content are freely available online, accessing the full articles requires a subscription to the Historical Archive. This ongoing project continuously adds content to SpringerLink, with more materials expected to be digitized throughout 2005. For further details about the Historical Archive project, you can reach out to your Springer licensing manager or visit our librarian page on springeronline.com.

Alternatively, if you prefer immediate access to a specific article, you have the option to purchase Historical Archive articles securely through our platform.

8.6. Pay-Per-View

8.6.1. Buying Articles on SpringerLink

To purchase individual articles electronically, begin by logging into SpringerLink using your SpringerLink username and password. Once logged in, you can either search for the desired article using the "Search" or "Browse" functions.

If you choose to search, enter keywords related to the article you're seeking in the search bar. If you prefer browsing, navigate through the publications to locate the journal and issue containing the article. Then, click on the article title to access its abstract page.

On the article abstract page, you'll find options to either add the article to your "Shopping Cart" or "Purchase Now". Select the appropriate option and proceed to complete the purchase on our secure server.

Upon successful purchase, a link to the article will be displayed on your screen. Additionally, if you're logged into your SpringerLink account, the article will be stored in your "Order History" for future reference.

8.7. Personalization

8.7.1. Setting Up Alerts for New Issues on SpringerLink

To stay informed about the latest research in your field of interest, take advantage of SpringerLink's Alerting services, which automatically notify you *via* email when new issues are published electronically. Here's how to set up your alerts:

1. **Accessing Alerts**: Once logged into your SpringerLink account, navigate to the SpringerLink blue banner toolbar and click on "Alerts".

2. **Choosing Alert Type**: On the "Alerts" page, you can set up either a "Table of Contents" alert or a "Keyword" alert. For new issue notifications, select "Table of Contents".

3. **Customizing Your Alerts**: Clicking on "Table of Contents" takes you to the "Alerts" management page, where you'll find options to personalize your alerts.

- **Publications**: View an alphabetical list of Springer journals and books. Check the box next to the titles you're interested in, then click "OK" to enable email alerts for those selections.

- **Subjects**: Browse publications organized by subjects. Similarly, select the titles of interest, then click "OK" to activate alerts.

- **Confirmation**: After making your selections, a confirmation page will allow you to review and confirm your choices.

Once your alerts are set up, you'll receive automatic email notifications whenever new issues are published electronically in the selected journals.

8.7.2. Editing Your Alerts on SpringerLink

If you need to modify or remove your current alerts, follow these steps:

1. **Accessing Alert Settings**: Log into your individual account on SpringerLink and click on the "Alerts" button on the blue banner toolbar.

2. **Selecting Table of Contents Alerts**: Choose "Table of Contents Alerts" from the options provided.

3. **Viewing Current Alerts**: On the TOC Alerting page, select the option "show only my current publications" and click on the "Publications" button.

4. **Managing Your Alerts**: You'll see a list of your current alert subscriptions. To modify or remove any alert, simply uncheck the box next to the title you wish to adjust.

5. **Confirming Changes**: After making your adjustments, click the "OK" button on the right-hand side of the page. A confirmation page will appear, allowing you to review the changes you've made.

By following these steps, you can easily edit your alerts to ensure you receive notifications tailored to your preferences.

8.7.3. Setting Up Favorites on SpringerLink

Creating "Favorites" on SpringerLink allows you to have quick access to your preferred journals directly from the homepage. Here's how to set up your favorites:

1. **Accessing Favorites**: After logging into your SpringerLink account, locate the "Favorites" button on the blue banner toolbar and click on it.

2. **Selecting Journals**: On the "Favorites" page, you'll see a list of Springer titles arranged alphabetically. Check the box next to the title of the journal you want to add as a favorite.

3. **Applying Changes**: Once you've selected your favorite journal(s), click the "Apply Changes" button on the right side of the "Favorites" page.

4. **Viewing Favorites**: Your selected favorite journal(s) will now appear beneath the "Login" box on the SpringerLink homepage for convenient access.

By following these steps, you can easily set up your favorites on SpringerLink, ensuring easy and quick access to your preferred journals.

8.7.4. For Assistance with Personalization Features

For inquiries regarding personalization features or any questions related to personalizing your experience on SpringerLink, please reach out to us at springerlink.marketing@springer.com. We're here to assist you with any queries you may have about optimizing your usage of our platform.

8.8. Printing and Downloading Articles

8.8.1. Trouble with Downloading or Printing? Here's How to Fix It.

To access the download or print functions for articles on SpringerLink, it's essential to utilize the Adobe Acrobat toolbar. Attempting to use the "File" option within your internet browser might not yield the desired results. Here's how you can proceed:

When using Adobe Acrobat, look for the download function, symbolized by a computer disk icon, and the print function, represented by a printer icon, both located on the Adobe Acrobat toolbar.

8.8.2. Troubleshooting: Saving HTML Full-Text Articles from SpringerLink

Saving HTML full-text articles from SpringerLink is typically straightforward, but certain systems may encounter issues with this process. If you're facing difficulties downloading HTML articles using the right-click function on your mouse, it could be due to compatibility issues with Internet Explorer 6.0 and specific Microsoft operating systems or service packs.

Microsoft provides an update that addresses this issue, applicable to Internet Explorer 6 Service Pack 1 (SP1) on various operating systems, including Windows XP, Windows Millennium Edition (Windows Me), Windows 2000, Windows 98, and Windows NT 4.0. Here are the corresponding download links:

For the English version:

http://www.microsoft.com/downloads/details.aspx?FamilyID=2F13B107C509-497C-B5F5-C86154DA4E91&displaylang=en

For the German version:

http://www.microsoft.com/downloads/details.aspx?displaylang=de&FamilyID=2f13b107-c509-497c-b5f5-c86154da4e91

Updating your Internet Explorer with this patch should resolve any issues you encounter when saving HTML articles from SpringerLink.

8.9. Accessing Publication Details on SpringerLink

Accessing Publication Details on SpringerLink: How to Navigate Journal Information If you're seeking specific publication details for titles on SpringerLink, such as Editorial Board information, Instructions for Authors, or Copyright Information, you can easily find them on the journal homepage. To begin, navigate to the SpringerLink toolbar and click on "Browse." From there, select your desired criteria—whether it's journals, books, or both—and click the "Browse" button.

Once you've selected a journal title, you'll be directed to its homepage. Here, you can access all pertinent publication details by clicking on the link labeled "About this Journal" (or "About this Series" for book series), typically located on the right side of the screen. Additionally, for further information, you can visit www.springeronline.com.

8.9.1 Rights and Permissions

8.9.1.1. Accessing Rights and Permissions Information for Springer Products

If you're in search of information regarding rights and permissions for Springer products, you can find comprehensive details on their corporate website, Springer Online

8.10. Viewing and Reading Articles

8.10.1. Accessing Full Text Articles for Subscribed Journals or Book Series

To access full text articles for your subscribed journals or book series on SpringerLink, simply look for the eyeglasses symbol next to the article. This symbol indicates that the full text of the article is available for viewing. You'll encounter this symbol in search results and on the issue level. SpringerLink provides two primary formats for viewing electronic articles: PDF or HTML. By either searching or browsing for a subscribed journal, you'll arrive at the publication's homepage, where volumes and issues available electronically are listed. Typically, you should have access to view all full texts of the journals to which you or your institution subscribe.

8.10.2. When Searching for an Article, Full Text Unavailable Appears

If you encounter the message "Full Text Unavailable" when viewing an article abstract page, it indicates that the full text of the article is not accessible. This situation primarily arises for articles published before 1997 when Springer initially introduced electronic publishing, starting with abstracts only. However, we are actively working to expand our full text archive, aiming to complete this project for content predating 1996 by the end of the current year.

8.10.3. I Need Access to an Article from 1978 but Can Only See the Abstract - What Should I Do?

For articles published before the digital era, our SpringerLink Historical Archive Project aims to digitize content from 1843 to 1996, previously available only in print. This endeavor now provides electronic access to previously unavailable material, marked as Historical Archive on separate journal homepages. While abstracts for all Historical Archive content are freely accessible online, full article access requires a subscription to the Historical Archive. Our ongoing project continuously adds content to SpringerLink, with updates expected throughout 2005. If you're interested in learning more about the Historical Archive project or obtaining access, please reach out to your Springer licensing manager. Additionally, like any other article on SpringerLink, Historical Archive articles can be purchased securely through our transaction system.

Section 6
Conclusion

CHAPTER 12

Conclusion: The Symphony of Life - Integrating Knowledge and Advancing Frontiers

As we reach the culmination of this comprehensive exploration into the biological sciences, it is essential to reflect on the interconnected themes and lessons presented throughout this book. The journey we embarked upon has traversed the fundamental principles of life, the complexities of ecosystems, the mechanisms driving evolution, and the frontiers of modern biological research and information systems. Each chapter has contributed to a holistic understanding of biology, painting a detailed portrait of the natural world and our place within it.

In the initial chapters, we laid the groundwork by delving into the structure and function of cells, the basic units of life. Understanding cellular components and their roles provided a foundation upon which more complex biological processes are built. The detailed examination of cellular reproduction, through mitosis and meiosis, highlighted the continuity of life and the intricate mechanisms that ensure genetic information is accurately passed on to the next generation.

Transitioning to genetics, we traced the historical evolution of genetic thought, from early hypotheses to the sophisticated molecular understanding we have today. By exploring the chemical foundations and structural intricacies of genes, we gained insight into how genetic information is encoded, transmitted, and expressed. This understanding is crucial for appreciating the profound impact genetics has on all living organisms, influencing traits, behaviors, and evolutionary trajectories.

The chapters on the diversity of life celebrated the vast array of organisms and their developmental processes. We journeyed through the origins of life, the ecological roles of fungi, and the detailed processes of animal development. The exploration of evolutionary forces underscored the dynamic nature of life, driven by natural selection, genetic drift, and gene flow. Understanding population dynamics and ecological interactions revealed the delicate balance that sustains ecosystems and the impact of human activities on this balance.

In the realm of advanced biological research, we explored the cutting-edge field of cancer systems biology, emphasizing the importance of a systems-level approach to understanding and treating this multifaceted disease. The evolution and future prospects of the ISI Web of Knowledge platform highlighted the revolution in accessing and analyzing scientific literature, enabling researchers to discover new connections and drive scientific progress.

The practical guide in Section V, "Crafting the Scholar's Path," equipped aspiring researchers with essential skills for navigating the academic manuscript journey. From crafting effective tables and figures to developing a robust first draft and selecting appropriate journals, this section provided invaluable guidance. Detailed guidelines for writing and submitting manuscripts, promoting publications, and utilizing language editing services were designed to support researchers in disseminating their findings effectively.

In Section VII, the comprehensive answers to questions and tasks reinforced key concepts and ensured a deep understanding of the material. This section served as a vital resource for self-assessment and further learning.

The appendices in Section VIII offered valuable supplementary materials, including general vocabulary lists and the fundamental components of medical English vocabulary. These resources aimed to support readers in mastering the specialized language of biology and medicine, enhancing their ability to engage with scientific literature and communicate their findings effectively.

The Road Ahead: Emerging Fields and Ethical Considerations

As we look to the future, the field of biological sciences continues to evolve, driven by advancements in technology and interdisciplinary approaches. Emerging fields such as synthetic biology, systems ecology, and bioinformatics promise to transform our understanding of life and address complex biological challenges. However, with these advancements come ethical and societal considerations. It is imperative to approach scientific progress with a sense of responsibility, ensuring that innovations benefit society and preserve the integrity of natural ecosystems.

The interconnectedness of life remains a central theme, highlighting the complexity and beauty of biological systems. By appreciating the intricate web of relationships that sustain life, we gain a deeper understanding of our role within this symphony of existence. As researchers, educators, students, and enthusiasts, we are called to contribute to the collective knowledge and stewardship of the living world.

Final Thoughts

This book aims to provide a thorough and accessible exploration of contemporary biological sciences. By bridging foundational knowledge with the latest research and technological developments, it offers readers a comprehensive understanding of the field. Whether you are a student beginning your journey, an educator seeking to inspire, a researcher pushing the boundaries of knowledge, or an enthusiast captivated by the wonders of life, this book serves as a valuable resource.

As we conclude this journey, we invite you to continue exploring, questioning, and discovering. The field of biology is ever-evolving, and the pursuit of knowledge is a lifelong endeavor. Embrace the curiosity that drives scientific inquiry, and let the beauty of the natural world inspire you to contribute to the advancement of biological sciences.

Thank you for embarking on this journey with me. May this book ignite your passion for biology and deepen your appreciation for the intricate and awe-inspiring tapestry of life.

Section 7
Answers

Answers

Answers (Lesson 1):

Exploring the Inner Workings of Cells: Understanding the Structure and Function of Cellular Components

Exc. 1:

1. Polysome - **G.** RNA and ribosomes

2. Pinocytosis - **N.** cell drinking

3. Exocytosis - **k.** expel

4. Plastid - **I.** in plants only

5. Golgi complex - **O.** packaging

6. Flagella - **M.** whiplike

7. Phagocytosis - **F.** engulfment

8. Lysosome - **B.** baglike structure

9. Basal body - **D.** where flagella grow

10. Chemotactic - **E.** toward or away from a chemical stimulus

11. Nucleus - **H.** weblike

12. Vacuole - **L.** packaging

13. Ribosome - **A.** protein synthesis

14. Cytoskeleton - **H.** weblike

15. Mitochondrion - **C.** power generator

Exc. 2:

1. False

2. False: Ribosomes are synthesized in the nucleolus, a structure within the nucleus of eukaryotic cells. They are formed from ribosomal RNA (rRNA) and proteins. Once assembled, ribosomes can be found in the cytoplasm or attached to the endoplasmic reticulum. So, they are not derived from the nucleoli but are produced there.

3. False

4. True

5. True

6. True

7. True

8. True

9. False

10. True

11. True

12. True

13. False

14. False

Exc. 3:

1. Engulfed

2. Golgi complex, Lysosomes

3. Lipofuscin granules, lipofuscin granules, aging

4. Mitochondria - chloroplasts

5. Microfilaments - intermediate filaments -microtubules

6. Actin

7. Chemotactic

8. Basal body

Exc. 4:

1. **E.** None of the above. Properties associated with processes of life are not solely attributed to any single component listed.

2. **A.** Serve as organelles involved in protein synthesis

3. **D.** All of the above

4. **B.** Nucleoli

5. **A.** Hydrolytic enzymes

6. **A.** Engulfing solid particles within vacuoles

7. **D.** Mitochondria: powerhouse organelles

8. **A.** Inside the mitochondrial matrix

9. **D.** Plastid

10. **B.** A polysome: a ribonucleoprotein complex

11. **C.** Serve as reservoirs for proteins, fats, and starch

12. **D.** A cytoskeleton structure

13. **C.** Compacted into a region known as the nucleoid

14. **B.** At the sites of ribosomes

15. **C.** A polysome

16. **B.** An amino acid sequence referred to as a signal peptide

17. C. In the Golgi apparatus

18. C. Utilizing a contractile vacuole

19. A. Lysosomal compartment

20. B. A mitochondrion

Lesson One: Understanding Bird Migration

<u>**Section 1:**</u>

1. **B**. Reasons behind bird migration

2. **C**. Feathers

3. B. Unchanging

4. B. They migrate instinctively

5. D. Enhanced desire to eat

6. **C**. Amount of light exposure

7. A. They provide insulation against cold weather

8. B. They regulate gland activity

9. C. Fluctuations in daylight length

10. D. Birds are observed migrating from birth without prior experience

<u>**Section 2:**</u>

1. **A**. Live collectively

2. **B**. Convert it into honey

3. **D**. Bottles or jars

4. **A**. The type of flower from which the nectar was gathered

5. **C**. Laying eggs

6. **D**. Foraging for food

7. **A**. Regulating hive temperature

8. **C**. As a food source

9. **C**. Collecting honey from hives

10. **D**. The variety of flowers visited by bees

Answers (Lesson 2):

Cell Division (Cellular Reproduction): Exploring the Intricacies of Mitosis and Meiosis

Exc. 1:

1. Cytokinesis - **M.** Division of cytoplasm

2. Synapsis - **K.** Crossing over

3. Histone - **O.** Positively charged protein

4. Mitosis - **J.** Cell division

5. Cell cycle - **I.** Sequence of cell growth and division

6. Chalone - **L.** Inhibit cell division

7. Spindle - **F.** Set of microtubules

8. Chromatid - **N.** Single chromosome copy

9. Nucleosome - **A.** DNA + histones

10. Diploid - **B.** Two sets

11. Meiosis - **H.** Gamete production

12. Cell plate - **G.** Plant division

13. Sex chromosomes - **C.** X and Y

14. Karyotype - **E.** Chromosome display

15. Haploid - **D.** One set

Exc. 2:

1. False	**2.** True	**3.** False	**4.** True	**5.** True [#]
6. True	**7.** True	**8.** True	**9.** False	**10.** False

5: Centrioles are indeed most pronounced in animal cells. These structures are involved in organizing the spindle fibers during cell division, particularly during mitosis. While some plant cells do contain structures called centrosomes, which function similarly to centrioles, they typically lack the classic centriole structure found in animal cells. Therefore, in terms of prominence and visibility, centrioles are more pronounced in animal cells.

Exc. 3:

1. Karyotype, autosomes, sex chromosomes

2. Synthesis, replicated, synthesized

3. Microtubule-organizing

4. Synapsis

5. Two, diploid

6. Four, haploid

7. Kinetochores

8. Centromere

Exc. 4:

1. **A** - S phase

2. **C** - A homologous pair

3. **C** - All chromosomes other than the sex chromosomes

4. **D** - G_1

5. **A** - 46 chromosomes

6. **B** - Homologous

7. **E** - Interphase

8. **B** - DNA is wound around histones

9. **C** - Chromatin

10. **C** - Chromatid

11. **C** - Prophase

12. **B** - Centromeres split and chromatids start to move apart

13. **E** - Telophase

14. **D** - The centriole

15. **B** - One copy of each chromosome

16. **C** - A reduction in the chromosome number

17. **C** - Synapsis

18. **A** - A ring of actin filaments pinching the cell in two

19. **B** - Homologous chromosomes exchange corresponding pieces of genetic information

20. **D** - All of the above

Reading Understanding

1. C. The origin of the brain and central nervous system

2. B. Coelenterate

3. A. Began

4. B. Backbones

5. A. The cerebellum

6. D. The continued growth of the mammalian brain

Answers (Lesson 3):

From Ancient Notions to Modern Insights: Unraveling the Foundations of Genetics

Exc. 1:

1. Dominant - **D.** Always expressed

2. Phenotype - **O.** Appearance

3. Allele - **E.** Alternative forms

4. Homozygous - **I.** Similar

5. Recessive - **C.** Nondominant

6. P1 - **B.** Parental

7. Dihybrid cross - **L.** Two characters

8. Law of segregation - **M.** Separate

9. Gene - **G.** Basic units of heredity

10. Pangenesis - **A.** Hippocrates

11. F2 - **F.** Grandchildren

12. Sex-linked - **N.** X or Y chromosome

13. Heterozygote - **H.** Different

14. Genotype - **K.** Total alleles

15. Germ plasm theory - **J.** Weismann

Exc. 2:

1. True	**2.** True	**3.** True	**4.** False	**5.** True
6. True	**7.** True	**8.** True	**9.** True	**10.** False

Exc. 3:

1. Incomplete dominance

2. Thomas Hunt Morgan

3. Dihybrid crosses

4. Carl Correns and Hugo de Vries

5. Punnett square

6. Alleles: Gene

7. Self-fertilizing

Exc. 4:

1. **B**. Pangenesis

2. **A**. Blending inheritance

3.   **C**. Produce offspring that are identical to the parent

4.   **D**. All of the above

5.   **B**. Bears dissimilar alleles for that trait

6.   **A**. Allele segregation occurs randomly during meiosis

7.   **B**. Distinct alleles of the same gene

8.   **D**. pp

9.   **B**. The recessive trait will not be observed in the F1 phenotypes

10. **D**. Alleles of different genes segregate randomly, and fertilization occurs at random

11. **D**. RrYy [#]

12. **A**. Forecast the likelihoods of various allele combinations

13. **B**. When both carry two different traits

14. **C**. Incomplete dominance

15. **C**. Chromosomal theory of inheritance

16. **D**. Fruit flies

17. **A**. Traits located on sex chromosomes

18. **B**. The failure of homologous chromosomes to segregate during mitosis or meiosis

19. **E**. XX

20. **A**. A recombinant type

#11: Let's analyze the given information:

A homozygous plant bearing round yellow seeds (RRYY)

A homozygous plant bearing wrinkled green seeds (rryy)

When these two plants are crossed, all the offspring having round yellow seeds indicates that the dominant alleles for both traits are present in the offspring.

The dominant allele for seed shape (round) is represented by R, and the dominant allele for seed color (yellow) is represented by Y.

Since all the offspring have round yellow seeds, they must inherit at least one dominant allele for each trait. Therefore, the genotype of the F1 generation would be RrYy.

So, the correct answer is d. RrYy.

Reading Understanding

1. **C.** How some insects camouflage themselves

2. **C.** Creatures that eat insects

3. **A.** By holding its body stiff and motionless

4. **A.** They resemble their surroundings all the time

5. **B.** Flowers

6. **B.** Paragraph two

Answers (Lesson 4):

The Chemical Foundations and Structural Unveiling of the Gene

<u>Exc. 1:</u>

1. 5' to 3' direction - **I**. DNA chain lengthens

2. Base - **D**. Ring structure composed of carbon and nitrogen

3. Chargaff's rules - **A**. A = T

4. Cytosine - **C**. A DNA base

5. Double helix - **E**. Shape proposed by Watson and Crick

6. Nucleoside - **F**. A base plus a sugar

7. Okazaki fragment - **H**. Small piece of DNA

8. Replication fork - **B**. Unwinding must occur

9. Semiconservative replication - **G**. Start of DNA replication

10. X-ray diffraction - **J**. Photographic process

<u>Exc. 2:</u>

1. True	**2.** True	**3.** False	**4.** False	**5.** True
6. False	**7.** False	**8.** True	**9.** True	**10.** False

11. True

Exc. 3:

1. Genes and enzymes (enzymatic function)

2. Electrophoresis

3. P. A. Levene

4. Nucleoside

5. Chargaff's rules or Chargaff

6. Template

7. Semiconservative

8. DNA ligase and polynucleotide ligase

Exc. 4:

1. **A**. Garrod

2. **C**. Separates molecules on the basis of their electric charge

3. **B**. The sickle cell gene chemically alters the hemoglobin protein

4. **A**. Deep red

5. **B**. One-gene-one-polypeptide hypothesis

6. **C**. Genes are made of nucleic acid

7. **C**. Virulent

8. **C**. A polysaccharide capsule was present on the S strain

9. **B**. DNA carries the genetic information

10. **C**. A nitrogenous base, a phosphate group, and a sugar molecule

11. **D**. Each nucleotide contains a different nitrogenous base

12. E. All of the above

13. E. A and C only

14. B. Watson and Crick

15. D. Pauling's work on DNA structure

16. B. Links the 5-C sugar of one nucleotide to the 3-C sugar of the next nucleotide

17. B. Weak hydrogen bonds

18. D. Each new molecule has one strand from the original molecule

19. E. One strand is oriented in a 5' to 3' direction and the other in a 3' to 5' direction

20. B. 5' to 3' direction

Reading Understanding

1. **C.** Three

2. **C.** How does water get to the tops of trees?

3. **B.** Showed

4. **A.** Some very tall trees have weak root pressure

5. **D.** Botanists have changed their theories of how water moves in plants

6. **D.** Evaporation

7. **B.** The attraction between water molecules is strong

Answers (Lesson 5):

Life's Story: From Start to Diversity

<u>**Exc. 1:**</u>

1. Stromatolite - **C**. Fossil marine algae

2. Coacervate - **M**. Polymer-rich spheres

3. Clade - **G**. Monophyletic group

4. Ozone layer - **L**. Filters out ultraviolet light

5. Pangaea - **K**. Single land mass

6. Crust - **D**. Outer rock skin

7. Species - **F**. First binomial name

8. Core - **A**. Composed of iron and nickel

9. Plate tectonics - **E**. Drifting of crust

10. Cladistics - **H**. Branching of closely related groups

11. Big bang - **N**. Origin of universe

12. Monophyletic - **I**. Share a common ancestor

13. Binomial system -**J**. Linnaeus

14. Mantle -**B**. The middle layer of the earth

<u>**Exc. 2:**</u>

1. False	**2.** True	**3.** True	**4.** True	**5.** False
6. False	**7.** True	**8.** False	**9.** True	**10.** True
11. True				

Exc. 3:

1. Continental drift

2. Paleozoic, mesozoic, and cenozoic

3. Laurasia and gondwana

4. Genus, species

5. Domain

6. Reproductive

7. Monophyletic

Exc. 4:

1. A. 18 billion years ago

2. A. Life arose spontaneously by chemical evolution

3. A. Monomer formation can occur

4. D. On clay or rock surfaces

5. E. All of the above

6. B. Between 3 and 4 billion years ago

7. B. By heating amino acids and exposing them to water

8. E. All of the above

9. C. 3.5 billion years old

10. C. Anaerobic and heterotrophic

11. C. Manufacture all their nutritional needs

12. A. Enabled cells to survive in shallow water and on land

13. B. The carbon cycle

14. A. 1.5 billion years ago

15. C. The mantle

16. A. A lipid bilayer

17. E. Precambrian

18. C. Two organisms cannot interbreed

19. A. Share a common ancestor

20. D. Cladistics

Reading Understanding

1. B. The phenomenon of leaf color transformation.

2. A. Amino acids.

3. B. Holding chlorophyll molecules.

4. C. Yellow or brown.

5. A. Relies on chlorophyll.

6. C. Which chemicals impart red and yellow hues to leaf tissue?

Answers (Lesson 6):

Fungi: The Mighty Decomposers

Exc. 1:

1. Rhizoid - **F**. "Fungus root"

2. Ergot - **N**. Source of drug LSD

3. Oomycetes - **D**. "Water molds"

4. Mycorrhiza - **H**. Analogous to plant root

5. Aerial hypha - **O**. Source of drug LSD

6. Conidium - **K**. Dustlike

7. Sporangium - **I**. Spore case

8. Basidiomycetes - **L**. Contains edible mushrooms

9. Lichens - **M**. Composite organisms

10. Haustorium - **C**. A fungal feeding structure

11. Dikaryon - **E**. A hyphal cell with two nuclei

12. Mycelium -**J**. Network of hyphae

13. Septate - **A**. Having cross walls

14. Ascus - **G**. Spore sac

Exc. 2:

1. **True** - Fungal mitosis indeed occurs within the nucleus, as mentioned in the text.

2. **False** - Survival spores are typically produced in smaller numbers, not necessarily large ones, as indicated in the text.

3. **True** - Oomycetes are usually diploid, as mentioned in the text.

4. **False** - Zygomycetes are typically haploid, not diploid, according to the text.

5. **False** - The statement is not mentioned in the text.

6. **False** - Oomycetes do not contain cellulose; they are known for their cell walls made of cellulose-like material, but not actual cellulose.

7. **True** - Mycorrhizal associations do occur among approximately 80 percent of land plants, as mentioned in the text.

8. **False** - Most basidiomycetes undergo sexual reproduction, not asexual reproduction, as stated in the text.

9. **True** - Crustose lichens do resemble tiny leaves, as indicated in the text.

10. **False** - Lichens are known for their ability to adapt to various climates, including dry climates, contradicting the statement.

11. **True** - The ratio of fungus to algae in lichens is approximately 9:1, as stated in the text.

Exc. 3:

1. Histones

2. Oomycetes

3. Mycorrhiza

4. 80

5. Budding

6. Basidiocarp

7. Pollution indicators

8. Chitin and, in some species, glucans

Exc. 4:

1. **C.** Through digestion outside their cells

2. **B.** Saprobes

3. **D.** Thallus

4. **A.** Feeding structures

5. **D.** Chitin

6. **B.** They have a multinucleate cytoplasm

7. D. Industrial filaments

8. A. Arises from the fusion of two genetically distinct individuals

9. C. Heterokaryosis

10. B. Short-lived

11. A. By producing millions of spores

12. E. Are all of the above

13. D. Their evolutionary relationships are unknown

14. D. Ophiostoma ulmi

15. B. Are terrestrial

16. D. Mycorrhizae

17. D. Basidiomycetes

18. C. Conidia

19. E. Ninety %

20. B. Their ability to absorb inorganic nutrients

Reading Understanding

I)   **1. B**. How trees endure and thrive   **2. B**. Light green   **3. A**. Departing

1.   **B.** Depict the network of food interactions among flora and fauna.

2.   **A.** A complex arrangement comprising multiple interconnected food chains.

3.   **D.** Plants like grass

4.   **B.** Their method of obtaining nourishment

5.   **C.** A plant

Answers (Lesson 7):

The Journey of Animal Development: From Gamete Production to Fertilization

Exc. 1:

1. Yolk - **I**. Food

2. Amnion - **L**. Cushions embryo

3. Regeneration - **O**. Replacement of lost parts

4. Parthenogenesis - **F**. "virgin" birth

5. Cleavage - **H**. Divides a single-celled zygote into many small cells

6. Zygote - **J**. Fertilized egg

7. Chorion - **N**. Fuses with allantois

8. Testis - **C**. Homologous with ovaries

9. Cortical reaction - **E**. Prevents multiple fertilizations

10. Ovum - **A**. Gamete or egg

11. Primitive streak - **K**. Associated with large yolks

12. Blastomere - **G**. Individual blastula cells

13. Allantois - **M**. "trash dump"

14. Oviduct - **D**. Egg tube

15. Gonial cell - **B**. Spermatogonia

Exc. 2:

1. **True** - Sertoli cells indeed play a supportive role in sperm production.

2. **False** - The sperm head contains very few or no mitochondria; they are mainly located in the midpiece and tail.

3. **True** - Oogenesis can indeed halt for extended periods in certain species, especially in mammals where oocytes remain arrested in prophase I of meiosis until ovulation.

4. **True** - An ostrich egg is a single cell and can be regarded as a singular cell entity.

5. **False** - During the "lamp brush" phase, chromosomes actively generate mRNA.

6. **True** - The initial contact of the sperm and egg triggers the acrosome reaction.

7. **False** - Before fertilization, the egg carries a negative electrical charge.

8. **False** - Parthenogenesis can result in the birth of both males and females, depending on the species.

9. **False** - The "animal pole" harbors the highest concentration of yolk. On the other hand, In organisms with minimal yolk, such as mammals, cleavage progresses uniformly, indicating that the yolk concentration is not highest at a specific pole.

10. **True** - The gut cavity indeed originates from an archenteron during gastrulation.

11. **False** - Amphioxus eggs do not exhibit a distinct primitive streak.

Exc. 3:

1. Sertoli

2. Golgi complex

3. Albumen

4. Jelly coat

5. Fertilization membrane

6. Morula

7. Morphogenesis

8. Neurulation

Exc. 4:

1. **C**. Seminiferous tubules

2. **A**. Mitochondria

3. **E.** Just after fertilization

4. **C**. Synthesize proteins and nucleic acids crucial for embryonic development

5. **D**. Mating behavior

6. **B**. Reaction of the egg's outer layer

7. **D**. Parthenogenesis

8. **B**. Yolk

9. **C**. Blastula

10. B. Cells are organized according to their future function

11. B. Gastrulation

12. C. Primitive streak

13. B. Skin

14. B. Differentiated

15. C. Cell shape

16. D. All of the above

17. D. The amount of food in the egg

18. E. Differentiation

19. C. Dedifferentiation solely transpires during regeneration.

20. D. Dedifferentiation

Reading Understanding

1. C	**2.** B	**3.** D	**4.** D
5. D	**6.** A	**7.** B	

Answers (Lesson 8):

Evolutionary Forces Shaping Species Diversity

<u>**Exc. 1:**</u>

1. Analogy - **H**. A functional comparison

2. Homology - **G**. Same ancestral origins

3. Microevolution -**D**. Minor allele frequency changes

4. Extinction - **J**. Total species death

5. Hybrid sterility - **I**. No one breeds mules

6. Macroevolution - **E**. Large phenotype changes

7. Species - **A**. Reproductively isolated

8. Allopatric speciation - **F**. Geographic separation

9. Polyploidization -**B**. Multiple chromosome sets

10. Phylogeny - **C**. Family tree

Exc. 2:

1. **False** - Temporal isolation functions as a prezygotic mechanism.

2. **True** - Hybrid sterility represents a postzygotic mechanism.

3. **False** - Mules are *sterile* and do not demonstrate heterozygote advantage.

4. **True** - Subspecies are geographically or ecologically distinct populations within a species and do not possess genetic superiority over full species.

5. **True** - The allopatric speciation model is rooted in geographic factors.

6. **False** - Autopolyploids emerge from the duplication of chromosomes within a single species, not hybridization.

7. **True** - Carbon-14 dating lacks reliability beyond 40,000 years.

8. **True** - A phylogeny delineates lineages of descent.

Exc. 3:

1. Convergent evolution

2. Parallel evolution

3. Analogous

4. Homologous

5. Extinction

6. Punctuated equilibrium

Exc. 4:

1. **D.** Reproductive barriers

2. **B.** Hybrid sterility

3. **A.** Ecological isolation

4. E. The establishment of a barrier to gene flow

5. C. Hybrid inviability

6. C. Polyploidization

7. B. Allopatric speciation

8. D. Regulatory genes

9. C. Sympatric speciation

10. D. Speciation in a population of grasshoppers

11. A. Differing environmental conditions

12. B. Phylogeny

13. B. Convergent evolution

14. D. The bird's wing and the fly's wing

Homology refers to similarity in structure due to common ancestry. Options a, b, c, and e all involve comparisons between structures of different species that share a common ancestor, indicating homology. However, the wings of birds and flies are not considered *homologous* despite their similar function in flight. Instead, they are considered *analogous* structures, as they have evolved independently in response to similar environmental pressures (convergent evolution) rather than from a shared ancestor.

15. B. Eldridge and Gould

16. A. Divergence

17. C. The slow rate of change in species over time

18. B. It does not apply to asexual organisms

19. B. The emergence of new ecological opportunities for a species

20. C. Polyploidy

Reading Understanding

I.

1. **B**   2. **C**   3. **D**   4. **D**

II.

1. **B**   2. **D**   3. **D**   4. **A**   5. **C**

Answers (Lesson 9):

Population Dynamics and Ecology

Exc. 1:

1. Natality - **B**. Per capita birth rate

2. K-selected species - **E**. Reproduction tied to k

3. Interspecific competition - **D**. Between species

4. Allelopathy - **G**. Chemical warfare

5. Population density - **H**. Individuals per area

6. Exponential growth - **I**. "no limits"

7. R-selected species - **C**. Reproduce many young

8. Mortality - **J**. Death rate

9. Carrying capacity - **A**. k

10. Intraspecific competition - **F**. Within species

Exc. 2:

1. **True** - The maximum sustainable population in an ecosystem, known as the carrying capacity, cannot be surpassed indefinitely without causing damage to the ecosystem.

2. **True** - Logistic growth results in a sigmoidal (S-shaped) curve as the population levels off when approaching the carrying capacity.

3. **False** - Species with r-selected traits typically lay many small eggs with few resources, and not substantial yolk reserves.

4. **False** - The distribution of a horse herd is typically clumped, not random.

5. **False** - Density-independent factors include natural disasters and weather conditions, not competition and predation, which are density-dependent factors.

6. **True** - Allelopathy involves the synthesis and release of chemicals that inhibit the growth of competing species.

7. **True** - Galapagos finches show distinct morphological differences, an example of character displacement, due to ecological competition.

8. **True** - Exponential growth is not sustainable over extended periods in natural populations because resources become limited.

Exc. 3:

1. Population density

2. Survivorship curve

3. Intraspecific

4. Reproductive time lag

5. K

6. Uniform

Exc. 4:

1. **B**. Rate

2. **D**. The host population is densely packed

3. **A**. Exponentially

4. **E**. Options b and c

5. **D**. A shift in the birth rate

6. **A**. The number of youthful individuals surpasses the number of elderly individuals

7. **D**. Commencement of reproduction at a comparatively advanced age

8. **A**. Interspecific competition

9. **D**. They are a K-selected species

10. **C**. Character displacement

11. **E**. All of the aforementioned

12. **E**. Cycling of the prey population and extinction of the predator

13. **C**. Allelopathy

14. **B**. Resource partitioning

15. **B**. S-shaped

16. **C**. 30 billion

17. **B**. Resources are abundant and patchy

18. **E**. All of the above

19. **B**. The community is simple and consists of very few species

20. **C**. A red snapper

Reading Understanding

I.

1. A   **2.** D   **3.** D   **4.** C   **5.**   A   **6.**   B   **7.**   C

II.

1. B   **2.** C   **3.** D   **4.** A   **5.** B   **6.**   A   **7.**   C

Answers (Lesson 10):

Navigating Cancer Systems Biology

<u>**Exc. 1:**</u>

1. **K.** Angiogenesis - Formation of new blood vessels.

2. **M.** Biomarker - Measurement and interpretation of biological markers.

3. **N.** Chromatin - Complex of DNA and protein in the nucleus.

4. **O.** DNA Methylation - Chemical modification of DNA affecting gene expression.

5. **L.** Enzyme-Linked Immunosorbent Assay (ELISA) - Laboratory method for detecting and

6. **I.** Genomic Instability - Measure of the stability and integrity of the genome.

7. **A.** Histone Modification - Addition of chemical groups to histone proteins affecting gene expression.

8. **H.** Immunohistochemistry - Use of antibodies to detect specific antigens in tissues.

9. **J.** Kinase - Enzyme adding phosphate groups to proteins.

10. **D.** MicroRNA - Short sequences of RNA regulating gene expression post-transcriptionally.

11. B. Oncogenesis - Process by which cancer forms.

12. G. PCR (Polymerase Chain Reaction) - Technique to amplify DNA sequences.

13. F. Signal Transduction - Uncontrolled division of cells leading to tumor development.

14. E. Telomere - Protective caps at the ends of chromosomes.

15. C. Transcriptomics - Analysis of RNA transcripts to study gene expression.

Exc. 2:

1. **True** - Angiogenesis refers to the formation of new blood vessels.

2. **False** - Biomarkers are used to measure and interpret biological markers, not specifically the stability of the genome.

3. **False** - Chromatin is composed of DNA and proteins found in the nucleus, not the cytoplasm.

4. **True** - DNA methylation can affect gene expression by adding chemical groups to DNA.

5. **False** - ELISA is a laboratory method for detecting and quantifying proteins, not for amplifying DNA sequences.

6. **True** - Genomic instability is a hallmark of many cancers.

7. **False** - Histone modification involves the addition of chemical groups to histone proteins, not DNA.

8. **True** - Immunohistochemistry uses antibodies to detect specific proteins in tissue samples.

9. **False** - Kinases are enzymes that add phosphate groups to proteins, not remove them.

10. **True** - MicroRNAs regulate gene expression by interacting with mRNA.

11. **True** - Oncogenesis is the process by which normal cells transform into cancer cells.

12. **False** - PCR (Polymerase Chain Reaction) is a technique used to amplify DNA sequences, not to study protein-protein interactions.

13. **True** - Signal transduction pathways are crucial for transmitting signals within a cell.

14. **True** - Telomeres are regions at the ends of chromosomes that protect them from degradation.

15. **True** - Transcriptomics involves the study of the complete set of RNA transcripts produced by the genome.

Exc. 3:

1. Blood vessels

2. Biological

3. Proteins

4. Suppress

5. Proteins

6. Mutations

7. Chemical

8. Proteins

9. Phosphate

10. mRNA

Exc. 4:

1. **B**. Angiogenesis

2. **C**. Detect and measure biological changes

3. **A**. Chromatin

4. **B**. Suppresses gene expression

5. **B**. Detect and quantify proteins

6. **B**. Increased rates of mutations

7. **A**. Chromatin structure and gene expression

8. **D**. Proteins

9. **B**. Add phosphate groups to proteins

10. **B**. mRNA

11. **C**. Stable gene expression

12. **A**. Wound healing

13. **D**. All of the above

14. **C.** DNA and proteins

15. **C.** DNA

16. **C.** Detect and quantify antigens

17. **C.** Both a and b

18. **D**. All of the above

19. **B**. Antibodies

20. **B**.Inhibit translation

Reading Understanding

1. B	**2.** C	**3.** D	**4.** C	**5.** B
6. B	**7.** D	**8.** C	**9.** B	**10.** B

Answers (Lesson 11):

The Evolution and Future Prospects of the ISI Web of Knowledge Platform

<u>Exc. 1:</u>

1. Bibliometrics – **E.** Statistical analysis of books and publications

2. Citation Index – **A.** Database for tracking citations between publications

3. Cross-content Searching – **J.** Comprehensive view across multiple content types

4. Hypernavigation – **C.** Navigation through hyperlinks in digital environments

5. Multidisciplinary Content – **F.** Spanning multiple academic disciplines

6. Probabilistic Search Engine – **G.** Use of probabilistic models for information retrieval

7. RoboLinks – **B.** Automated linking system for full-text resources

8. Scientometrics – **H.** Analysis and quantification of scientific literature

9. Web of Science – **I.** Multidisciplinary citation database

10. Web of Knowledge – **D.** Integrated research platform combining multiple databases

<u>Exc. 2:</u>

1. **False** (Garfield introduced the concept in 1955)

2. **False** (It was initially available in print format)

3. **True**

4. **True**

5. **False** (It supports multidisciplinary research)

6. **True**

7. **False** (RoboLinks verifies links before displaying them)

8. **False** (It includes tools like ISI Journal Citation Reports and Essential Science Indicators)

9. **True**

10. **True**

11. **True**

12. **True**

13. **True**

14. **False** (It plans to expand into additional disciplines)

Exc. 3:

1. 1997

2. Citation

3. Science

4. Hyper

5. Essential

6. Biological

7. Multidisciplinary

8. Context

9. Internal, external

10. Full-text

<u>Exc. 4:</u>

1. **C**. 1997

2. **C**. Eugene Garfield

3. **C**. Science citation index

4. **B**. Eugene Garfield

5. **C**. Biological sciences

6. **B**. Multidisciplinary research

7. **B**. Hypernavigation

8. **B**. Performance metrics

9. **B**. Generate context-sensitive links

10. **C**. Unify access to both internal and external resources

Reading Understanding

A.

1. B	2. B	3. B	4. B	5. C
6. A	7. B	8. B	9. A	10. B

B.

1. B	2. B	3. B	4. B	5. B
6. B	7. B	8. C	9. C	10. B
11. B	12. B	13. B	14. B	15. B

Section 8
Appendix

Appendix: Vocabulary Prefixes and Suffixes Roots

Appendix- The fundamental components of medical English vocabulary: roots, prefixes, and suffixes commonly used.

1. Position-related roots		
Caud- (related to the Tail, in anatomy towards the lower part of the body)		**Later**- (related to Lateral, pertaining to the side)
Cephal- (related to Head)		**Medi**- (related to Medial, pertaining to the middle)
Coron- (related to Crown, top of the head)		**Mes**- (related to Mesial, middle)
Crani- (related to Skull)		**Pariet**- (related to Parietal, pertaining to the walls of the body)
Dextr- (related to Right side)		**Proxim**- (related to Proximal, near, closer to the point of reference)
Dist- (related to Far, farthest point from the center)		**Sinistr**- (related to Sinister, pertaining to the left side)
Extern- (related to External, outside)		**Transvers**- (related to Transverse, horizontal)
Front- (related to Front, forehead)		**Ventr**- (related to Ventral, pertaining to the abdomen or belly side)
Intern- (related to Internal, inside)		**Viscer**- (related to Visceral, pertaining to internal organs)

(Table) cont.....

2. Skeletal system-related roots

Arthr- (related to Joint)	**Patell**- (related to Patella: kneecap)
Articul- (related to Joint)	**Pector**- (related to Pectoral: chest)
Astragal- (related to Astragalus, bone in the ankle)	**Pelv**- (related to Pelvis: hip bone)
Burs- (related to Bursa, fluid-filled sac in joints)	**Phalang**- (related to Phalanx: finger or toe bone)
Calcane- (related to Calcaneus, heel bone)	**Por**- (related to Pore or Passage)
Carp- (related to Wrist)	**Pub**- (related to Pubic: pubic bone)
Cervic- (related to Neck)	**Rachi**- (related to Rachi: spine)
Chondr- (related to Cartilage)	**Ment**- (related to Mental: chin)
Clavic- (related to Clavicle)	**Myel**- (related to Myel: marrow or spinal cord)
Cox- (related to Hip)	**Nas**- (related to Nasal: nose)
Dactyl- (related to Finger or Toe)	**Radi**- (related to Radius: forearm bone)
Desm- (related to Ligament)	**Scapul**- (related to Scapula: shoulder blade)
Femor- (related to Femur, thigh bone)	**Scler**- (related to Scler: hardness or sclera)
Fibul- (related to Fibula, bone in the lower leg)	**Scoli**- (related to Scoliosis: lateral curvature of the spine)
Fleet/flex- (related to Flexibility)	**Spin**- (related to Spine)

(Table) cont.....

Gnath- (related to Jaw)	**Spondyl**- (related to Spondyl: vertebra)
Gon(at)/gony- (related to Generation, Birth)	**Stern**- (related to Sternum: breastbone)
Humer- (related to Humerus, bone in the upper arm)	**Tal**- (related to Talus: ankle bone)
Ischi- (related to Ischium, bone in the pelvis)	**Tars**- (related to Tarsus: ankle bones)
Lumb- (related to Lumbar, lower back)	**Tempor**- (related to Temporal: temple)
Mandibul- (related to Mandible, jawbone)	**Thorac**- (related to Thoracic: chest)
Maxill- (related to Maxilla, upper jawbone)	**Tibi**- (related to Tibia: shinbone)
Oss(e)- (related to Bone)	**Uln**- (related to Ulna: forearm bone)
Ost(e)- (related to Bone)	**Vertebr**- (related to Vertebra: backbone)

3. Roots related to the muscular and connective tissue	
Adip- (related to Fat, Adipose tissue)	**K/cin**- (related to Movement)
Coll(a)- (related to Collagen)	**K/cine(s)**- (related to Movement)
Cut- (related to Skin)	**Muc**- (related to Mucus)
Cyt- (related to Cell)	**Muscul**- (related to Muscle)
Derm(at)- (related to Skin)	**My**- (related to Muscle)

(Table) cont.....

Fasci- (related to Fascia)	**Sarc**- (related to Flesh, Muscle)
Fibr- (related to Fiber)	**Stear/t**- (related to Fat)
Hist- (related to Tissue)	**Syndesm**- (related to Ligament)
Hyal- (related to Hyaline)	**Ten(ont)**- (related to Tendon)
Hydr- (related to Water)	**Tend(in)**- (related to Tendon)
Hygr- (related to Moisture)	**Ton**- (related to Tonus, Muscle tone)
In- (related to Within)	

4. Nervous and sensory systems-related routes	
Aur- (related to Ear)	**Neur**- (related to Nerve)
Blephar- (related to Eyelid)	**Ocul**- (related to Eye)
Cerebell- (related to Cerebellum)	**Ophthalm**- (related to Ophthalmology)
Cerebr- (related to Cerebrum)	**Ot**- (related to Ear)
Conjunctiv- (related to Conjunctiva)	**Pupill**- (related to Pupil)
Cor(e)- (related to Heart)	**Scler**- (related to Sclera)
Corne- (related to Cornea)	**Thalam**- (related to Thalamus)
Cyst- (related to Cyst)	**Tympan**- (related to Tympanic Membrane)
Dacry- (related to Tear)	**Ventricul**- (related to Ventricle)
Encephal- (related to Brain)	

5. Cardiovascular-related routes

Aden- (related to Gland)	**Papill**- (related to Nipple or Papilla)
Angi- (related to Vessel)	**Phag**- (to eat, ingest)
Aort(ic)- (related to Aorta)	**Phleb**- (related to Vein)
Arteri- (related to Artery)	**Piez**- (pressure)
Atri- (related to Atrium)	**Plasm**- (formation or growth)
Brady- (slow)	**Poikil**- (variable, irregular)
Capill- (related to Capillary)	**Plasm**- (plasma)
Cardi- (related to Heart)	**Pneum(at)**- (related to Lung)
Drepan- (related to Sickle)	**Rhythm**- (rhythm, periodicity)
Erythr- (related to Red)	**Sangui(n)**- (related to Blood)
Ger / gest- (carry)	**Sept**- (partition)
Granul- (related to Granule)	**Ser**- (serum, related to blood)
Hem(at)- (related to Blood)	**Sphygm**- (pulse, related to heartbeat)
Lien- (related to Spleen)	**Splen**- (spleen)
Lymph(at)- (related to Lymphatic System)	**Tachy**- (fast, rapid)
Olig- (few, scanty)	**Thromb**- (related to Blood clot)
Pan(t)- (all, complete)	**Valv**- (valve)
Phleb- (related to Vein)	**Vas**- (vessel)

Piez- (pressure)	Ven- (vein)

6. Respiratory system-related routs

Aer- (related to Air)	**Pleur**- (related to Pleura)
Branch- (related to Branch)	**Phon**- (related to Sound)
Bronchi- (related to Bronchi)	**Phren**- (related to Diaphragm)
Cili- (related to Cilia)	**Pneum(on)**- (related to Lung)
Coni- (related to Dust)	**Pulm(on)**- (related to Lung)
Laryng- (related to Larynx)	**Rhin**- (related to Nose)
Nas- (related to Nose)	**Sin**- (related to Sinus)
Or- (related to Mouth)	**Trache**- (related to Trachea)
Ox(y)- (related to Oxygen)	

7. Digestive system-related roots

Abdomin- (related to Abdomen)	**Lingu**- (related to Tongue)
Bil- (related to Bile)	**Lith**- (related to Stone)
Blenn- (related to Mucus)	**Necr**- (related to Death)
Brom(at)- (related to Smell, *e.g.*, Bromopnea)	**Odont**- (related to Tooth)
Bucc- (related to Cheek)	**Palat**- (related to Palate)
C(a)ec- (related to Cecum)	**Pancreat**- (related to Pancreas)

(Table) cont.....

Celi- (related to Abdomen)	**Pancreatic**- (related to Pancreas)
Ch(e)- (related to Bile)	**Peps/ pept**- (related to Digestion)
Chol(e)- (related to Bile)	**Periton**- (related to Peritoneum)
Choledoch- (related to Common Bile Duct)	**Pharyng**- (related to Pharynx)
Col- (related to Colon)	**Ptyal**- (related to Saliva)
Copr- (related to Feces)	**Pyle**- (related to Gate, *e.g.*, Pylorus)
Dent- (related to Tooth)	**Pylor**- (related to Pylorus)
Duoden- (related to Duodenum)	**Rect**- (related to Rectum)
Enter- (related to Intestine)	**Sial**- (related to Saliva)
Esophag- (related to Esophagus)	**Sterc(or)**- (related to Feces)
Gastr- (related to Stomach)	**Stom(at)**- (related to Mouth)
Gingiv- (related to Gums)	**Uran**- (related to Palate)
Gloss- (related to Tongue)	**Uvul**- (related to Uvula)
Hepatic- (related to Liver)	**Xer**- (related to Dryness)
Lapar- (related to Abdomen)	

8. Roots related to the reproductive and urinary systems

Andr- (related to Man, Male)	**Orch(i/id)**- (related to Testicle)
Balan- (related to Gland, specifically the glans penis)	**Osche**- (related to Scrotum)

Cervic- (related to Neck, Cervix)	**Ov**- (related to Egg)
Colp- (related to Vagina)	**Ovari**- (related to Ovary)
Crypt- (related to Hidden, Crypt)	**Pen**- (related to Penis)
Epididym- (related to Epididymis)	**Phall**- (related to Phallus)
Episi- (related to Childbirth, Episiotomy)	**Prostat**- (related to Prostate)
Erot- (related to Sexual love, Eros)	**Pyel**- (related to Renal Pelvis)
Gon- (related to Offspring, Birth)	**Ren**- (related to Kidney)
Gonad- (related to Reproductive glands)	**Salping**- (related to Fallopian Tube)
Gravid- (related to Pregnant, Pregnancy)	**Semin**- (related to Semen)
Gyn(e/ec)- (related to Female, Women)	**Sperm(at)**- (related to Sperm)
Hyster- (related to Uterus, Hyster)	**Toc**- (related to Childbirth)
Meat- (related to Food, Meat)	**Ur**- (related to Urine)
Men- (related to Menses, Menstruation)	**Uretero**- (related to Ureter)
Metr- (related to Uterus, Metra)	**Urethr**- (related to Urethra)
Nephr- (related to Kidney)	**Vas**- (related to Vessel, specifically Vas Deferens)
Omphal- (related to Umbilicus, Navel)	**Vesic**- (related to Bladder)
OO- (related to Egg)	**Vir**- (related to Man, Male)

(Table) cont.....

Oophor- (related to Ovary)	

9. Endocrine and cardiovascular-related routes

Adren- (related to the Gland, specifically the Adrenal gland)	**Lact**- (related to Milk)
Areol- (related to Small space, Areola: pigmented area of skin around the nipple)	**Mamm**- (related to Breast, Mammary gland)
Crin- (related to Secretion, Production)	**Mast**- (related to Breast, Mammary gland)
Duc(t)- (related to Duct, Tube)	**Pineal**- (related to Pineal gland)
Galact- (related to Milk)	**Seb**- (related to Grease, Sebum)
Hidr- (related to Sweat)	**Thel(e)**- (related to the Nipple, specifically the Mammary gland)
Hormon- (related to Hormone)	**Thym**- (related to Thymus gland)
Hypophys- (related to Hypophysis, Pituitary gland)	**Thyr**- (related to Thyroid gland)
Kerat- (related to Horny tissue, Keratin)	

10. Microbes and parasites-related roots

Acar- (related to Mite, Acarid)	**Gam**- (related to the Gamete, population, for example, "Gametocyte" referring to the sexual cells of parasites)

(Table) cont.....

Actin- (related to Ray, Actin)	Myc(et)- (related to Fungus, Mycete)
Ameb- (related to Amoeba)	Myi- (related to Fly, for example, "Myiasis" which refers to diseases caused by the presence of flies)
Bacill- (related to Bacteria, Bacillus)	Phyt- (related to Plant)
Bacteri- (related to Bacteria)	Spor- (related to Spore)
Cocc- (related to Cocci, for example, "Coccidia" referring to a genus of single-celled parasites)	

11. Quantity-related roots	
Deca- (related to Ten)	**Mon-/uni- (related to One, Single)**
Deci- (related to Tenth)	Noni- (related to No, None)*
Di/bi- (related to Two)	Oct(a)- (related to Eight)
Dipl- (related to Double, Two-fold)	Penta(a)- (related to Five)
Ennea- (related to Nine)*	Primi- (related to First)
Hapl- (related to Single)	Quinti- (related to Fifth)
Hect-/cent- (related to Hundred)	Secundi- (related to Second)
Hemi-/semi- (related to Half)	Terti- (related to Third)
Hept(a)-/sept- (related to Seven)	Tetra-/quadri- (related to Four)
Hex(a)-/sex(ti)- (related to Six)	Tri- (related to Three)
Kil-/milli- (related to Thousand)	

* The prefixes "Ennea-" and "Noni-" both refer to numbers, but they represent different numerical values:

• **Ennea-** : This prefix denotes the number nine. It is derived from the Greek word "ennea," which means "nine." For example, "Enneadecagon" refers to a polygon with nine sides.

• **Noni-** : This prefix also denotes the number nine. It is derived from the Latin word "nonus," which means "nine." For example, "Nonagon" refers to a polygon with nine sides.

Both prefixes essentially mean the same thing, representing the number nine, but they come from different linguistic origins: **"Ennea-" from Greek** and **"Noni-" from Latin**.

12. The other roots

Amyl- (related to Starch)	**Hidr-** (related to Sweat)
Ankyl- (related to Joint, Stiffness)	**Hom-** (related to Similarity, Equality)
Bar- (related to Pressure)	**Homo**(o)e-/Homoi- (related to Same, Similar)
Brachi- (related to Arm)	**Pachy-** (related to Thickness)
Brachy- (related to Short)	**Ped-** (related to Foot)
Ch(e)ir- (related to Hand)	**Pil-** (related to Hair)
Derm(at)- (related to Skin)	**Pseud-** (related to False)
Dynam- (related to Power, Energy)	**Som(at)-** (related to Body)
Ger(ont)- (related to Old age, Aging)	**Thanat-** (related to Death)

13. Other common roots	Examples
A/ah- (related to Without)	• **Abnormal** • **Abortion**
A/an- (related to Without)	• **Analgesia Aseptic**
Actino- (related to Ray, Radiation)	• **Actinogram** • **Actinotherapy**
Aden- (related to Gland)	• **Adenoma** • **Adenotomy**
Adip- (related to Fat)	• **Adipoma** • **Adiposuria**
Ambi- (related to Both sides)	• **Ambilateral** • **Ambisexual**
Amphi/o- (related to Both, Double)	• **Amphibia** • **Amphigony**
Amylo- (related to Starch)	• **Amyloidosis** • **Amylomaltose**
An- (related to Without consciousness, Unconsciousness)	• **Anesthesia** • **Anesthetic**
Anis- (related to Unequal, Uneven)	• **Anisocoria** • **Anisometropia**

Angio- (related to Vessel, Blood vessel)	• **Angiotensin** • **Angiotonica**
Ante- (related to Before, Forward)	• **Anteoarchium** • **Antelabium**
Aqua/i- (related to Water)	• **Aquaeductus** • **Aquiform**
Archi- (related to First, Ancient)	• **Archicenter** • **Archigaster**
Arteri(o)- (related to Artery)	• **Arteriosclerosis** • **Artery**
Arthr- (related to Joint)	• **Arthralgia** • **Arthritis**
Atropho- (related to Wasting away, Atrophy)	• **Atrophoderma** • **Atrophy**
Audi/apui- (related to Hearing)	• **Audiphone** • **Auditorium**
Aur(ic)- (related to Ear)	• **Auricula** • **Auriscope**
Bacteri(o)- (related to Bacteria)	• **Bactericide** • **Bacteriolysis**

(Table) cont.....

Baso- (related to Base)	• **Basophilia**
Bili- (related to Bile)	• **Biliousness** • **Bilirubin**
Blast- (related to Germ, Bud)	• **Blastoderm** • **Blastocyst**
Blepharo- (related to Eyelid)	• **Blepharotomy**
Bronch- (related to Bronchus)	• **Bronchitis** • **Bronchopneumonia**
Calci- (related to Calcium)	• **Calcification**
Cancer- (related to Cancer)	• **Canceration** • **Canceremia**
Card(i/io)- (related to Heart)	• **Cardiotonics**
Cata- (related to Down, Under)	• **Catalysis**
Celio- (related to Abdomen)	• **Celiorrhaphy** • **Celiotomy**
Cephal- (related to Head)	• **Cephalalgia** • **Cephalocele**
Cereh(ro)- (related to Brain)	• **Cerebration** • **Cerebritis**
Cheil- (related to Lip)	• **Cheilalgia**

(Table) cont.....

	• **Cheilotomy** • **Cheilitis** • **Cheiloplasty**
Chemo- (related to Chemistry, Chemical)	• **Chemo-antigen** • **Chemophysiology** • **Chemotherapy**
Chol(e/o)- (related to Bile)	• **Cholemia** • **Cholochrome** • **Cholecystitis** • **Cholelithiasis** • **Cholestasis**
Cholecysto- (related to Gallbladder)	• **Cholecystalgia** • **Cholecystotomy**
Chondr- (related to Cartilage)	• **Chondritis** • **Chondroma** • **Chondrocyte** • **Chondroplasty**
Chromato/chrome- (related to Color)	• **Chromidrosis** • **Chromosome** • **Chromatography**

(Table) cont.....

	• **Chromatosis**
Co- (related to Together)	• **Coagulation** • **Coitus**
Con- (related to With)	• **Conception** • **Congestion**
Contra- (related to Against, Opposite)	• **Contraceptive** • **Contralateral**
Cortico- (related to Cortex)	• **Corticoid** • **Corticosterone** • **Corticospinal**
Cyesio- (related to Pregnancy)	• **Cyesiognosis** • **Cyesiology** • **Cyesiography** • **Cyesiometry** • **Cyesiopathy** • **Cyesiosynthesis**
Cysti/o- (related to Cyst, Bladder)	• **Cystitis** • **Cystocele** • **Cystography** • **Cystectomy**

(Table) cont.....

Cyto- (related to Cell)	• **Cytology**
	• **Cytoplasm**
	• **Cytokinesis**
	• **Cytochemistry**
	• **Cytogenesis**
	• **Cytotoxic**
De- (related to Remove)	• **Decomposition**
	• **Dedentition**
Dent(i)- (related to Tooth)	• **Dentalgia**
	• **Dentition**
	• **Dentifrice**
	• **Dentin**
	• **Dentist**
	• **Dentistry**
Dermat/dermo- (related to Skin)	• **Dermatitis**
	• **Dermatosis**
	• **Dermatome**
	• **Dermoplasty**
Des/dis- (related to Remove, Reduce, Separate)	• **Desensitize**
	• **Desquamation**

Dextro- (related to Right)	• **Dextrococaine** • **Dextrotorsion**
Dia- (related to Through, Across)	• **Diathermia** • **Dialysis**
Dipl- (related to Double, Two-fold)	• **Diplococus** • **Diplocephaly**
Dis- (related to Not, Apart, Away)	• **Discharge** • **Disfunction**
Dys- (related to Bad, Difficult, Abnormal)	• **Dyspepsia** • **Dysuria**
Ecto- (related to External, Outer)	• **Ectoparasite** • **Ectoderm**
Encephalo- (related to Brain)	• **Encephalalgia** • **Encephalorrhagia**
End- (related to Inner, Within)	• **Endocardium** • **Endocrine**
Enter- (related to Intestine)	• **Enteritis** • **Enterostenosis**
Ento- (related to Inner, Within)	• **Entoderm** • **Entocyte**

(Table) cont.....

	• **Entocornea**
Epi- (related to Upon, Above)	• **Epidermis** • **Epigastrium**
Eu- (related to Good, Normal)	• **Euphoria** • **Eutocia** • **Euthanasia**
Ex- (related to Out of, Beyond)	• **Excoriation** • **Exhale** • **Expectoration**
Exo- (related to External, Outer)	• **Exocardia** • **Exocoelom** • **Exoskeleton** • **Exoplanet**
Extra- (related to Beyond, Extra)	• **Extracurricular** • **Extra-oral** • **Extraterrestrial** • **Extravasation**
Febri- (related to Fever, Body heat)	• **Febricula** • **Febrifuga** • **Febrile**

Ferri- (related to Iron or containing Iron)	• Ferrigluconate • Ferritin • Ferrioxide
Ferro- (related to Iron)	• Ferrohemoglobin • Ferrobacillus • Ferrous
Fibr/i/o- (related to Fibers or Tissue)	• Fibrination • Fibroadenoma • Fibroblast • Fibrosis
Flav- (related to Yellow, associated with the color yellow)	• Flavonoid • Flavonone • Flavoxanthin
Follicul- (related to Follicle)	• Folliculitis • Folliculosis
Galact- (related to Milk)	• Galactemia • Galactometer • Galactose • Galactopoiesis
Gastr- (related to Stomach)	• Gastrectomy

(Table) cont.....

	• **Gastric bypass**
	• **Gastritis**
	• **Gastroenterology**
	• **Gastrosis**
	• **Gastrectasis**
Genito- (related to Genital)	• **Genitoplasty**
	• **Genitourinary**
	• **Genitalia**
Giga/ giganto- (related to Giant, Gigantic)	• **Gigantocyte**
	• **Gigantosoma**
Glob- (related to Globulin or Protein)	• **Globulinuria**
	• **Globulolysis**
Gluco/glyco- (related to Sugar, Glucose)	• **Glucose**
	• **Glycosuria**
Gonado- (related to Gonads, Sexual Glands)	• **Gonadoectomy**
	• **Gonadogenesis**
Granulo- (related to Granules)	• **Granuloma**
	• **Granulopenia**
Gyn/gynae(co)- (related to Woman, pertaining to women)	• **Gynaecology**
	• **Gynopathia**

Haem(ato)- (related to Blood)	• **Hemataemesis** • **Hemochrome**
Helio- (related to Sun)	• **Heliosis** • **Heliotherapy**
Hepat(ico)- (related to Liver)	• **Hepatitis** • **Hepatomegaly**
Heredo- (related to Hereditary)	• **Heredo-immunity** • **Heredopathia**
Hernio- (related to Hernia)	• **Hernioplasty** • **Herniopuncture**
Hist(io)- (related to Tissue)	• **Histamine** • **Histiocytoma**
Holo- (related to Whole, Entire)	• **Holoenzyme** • **Holotonia**
Hom(e)o/homoeo- (related to Similar, Similarity)	• **Homeochronous** • **Homeostasis** • **Homoxygote** • **Homeopathy**
Hydro- (related to Water)	• **Hydronephrosis** • **Hydrophobia**

Hyster(o)- (related to Uterus)	• **Hysterectomy** • **Hysteritis** • **Hysteroscopy**
Iatro- (related to Physician, Treatment)	• **Iatrochemia** • **Iatrica** • **Iatrogenic** • **Iatrology**
Ile- (related to Ileum, part of the small intestine)	• **Ileitis** • **Ileostomy** • **Ileum**
Ilio- (related to Ilium, part of the hip bone)	• **Iliolumbar** • **Ilium**
Infra- (related to Below, Beneath)	• **Infraclavicular** • **Infrasonic** • **Infracostal** • **Inframaxillary**
Ischio- (related to Ischium, part of the hip bone)	• **Ischiopagus** • **Ischiorectal** • **Ischiopagus** • **Ischium**

Iso- (related to Equal, Same)	• **Isometric** • **Isothermal**
Jejuno- (related to Jejunum, the middle part of the small intestine)	• **Jejunostomy** • **Jejunitis**
Kera(to)- (related to Cornea)	• **Keratitis** • **Keratoplasty**
Labio- (related to Lip)	• **Labioplasty** • **Labiomental**
Lacri- (related to Tears, Lacrimal gland)	• **Lacrimation** • **Lacrimator** • **Lacrimal**
Lacto- (related to Milk)	• **Lactation** • **Lactacidemia** • **Lactose**
Laparo- (related to Abdomen)	• **Laparorrhaphy** • **Laparoscopy** • **Laparotomy**
Laryng- (related to Larynx)	• **Laryngology** • **Laryngopathy** • **Laryngitis**

(Table) cont.....

	• **Laryngoscopy**
Lip(o)- (related to Fat)	• **Liposuction** • **Lipoma**
Lith- (related to Stone)	• **Lithiasis** • **Lithuresia** • **Lithotripsy** • **Lithogenesis**
Lumbo- (related to Loin, Lower back)	• **Lumbodynia** • **Lumbosacral**
Lymph- (related to Lymph)	• **Lymphocyte** • **Lymphatic** • **Lymphoidectomy**
Lys- (related to Breakdown, Destruction)	• **Lysinogen** • **Lysis** • **Lysozyme**
Mal- (related to Bad, Inappropriate)	• **Malformation** • **Maldigestion** • **Malabsorption**
Mammilli- (related to Mammilla, Nipple)	• **Mamillitis** • **Mamilliform**

Mammo- (related to Breast)	• **Mammalia** • **Mammectomy** • **Mammography** • **Mammoplasty**
Mast- (related to Breast)	• **Mastadenitis** • **Mastitis** • **Mastectomy**
Medico- (related to Medicine, Medical)	• **Medicogalyanic** • **Medico-legal** • **Medico-social**
Mega(lo)- (related to Large)	• **Megaglobulinaemia** • **Megalokaryocyte** • **Megalomania** • **Megacolon**
Mela(no)- (related to Black, related to skin pigmentation)	• **Melanoma** • **Melanocyte** • **Melana** • **Melasma**
Meningo- (related to Meninges, the membranes covering the brain and spinal cord)	• **Meningitis** • **Meningioma**

(Table) cont.....

	• **Meningorrhagia**
Meso- (related to Middle, Intermediate)	• **Mesoderm**
	• **Mesophyloma**
Meta- (related to Change, Beyond)	• **Metabolism**
	• **Metamorphosis**
	• **Metanephros**
	• **Metastasis**
Metro- (related to Uterus)	• **Metritis**
	• **Metroplasty**
	• **Metrosalpingitis**
Morpho- (related to Shape, Structure)	• **Morphodifferentiation**
	• **Morphology**
	• **Morphogenesis**
	• **Morphogeny**
Mort- (related to Death)	• **Mortality**
	• **Mortuary**
Muci/o- (related to Mucus)	• **Muciform**
	• **Mucocutaneous**
Musculo- (related to Muscle)	• **Musculocutaneous**
	• **Musculoskeletal Musculature**

My(o)- (related to Muscle)	• **Myocardium** • **Myopathy**
Myelo- (related to Spinal cord, Bone marrow)	• **Myeloid** • **Myelopathy** • **Myelorrhagia** • **Myelotomy**
Narco- (related to Sleep or Numbness)	• **Narcodiagnosis** • **Narcolepsy** • **Narcosis** • **Narcotic**
Neo- (related to New)	• **Neoplasm** • **Neonatal**
Nephr- (related to Kidney)	• **Nephrectomy** • **Nephritis**
Nerv- (related to Nerve)	• **Neuralgia** • **Nerve-centre** • **Nervine**
Neur- (related to the Nerve or Nervous System)	• **Neurogenesis** • **Neuropathy**
Neutro- (related to Neutrophils or neutral)	• **Neutrality**

(Table) cont.....

	• **Neutralization**
Normo- (related to Normal)	• **Normoglycemia** • **Normocyte** • **Normotension**
Nov- (related to New)	• **Novarsenol** • **Novaspirin**
Nucleo- (related to Nucleus)	• **Nucleotide** • **Nucleolus** • **Nucleoprotein**
Oculo- (related to Eye)	• **Oculogyration** • **Oculopathy**
Odonto- (related to Tooth)	• **Odontobothrion** • **Odontolith**
Oo- (related to Egg, Ovum)	• **Oocyte** • **Oogenesis** • **Oothectomy**
Ophthalm- (related to Eye)	• **Ophthalmectomy** • **Ophthalmologist** • **Ophthalmology**
Orchi(do)- (related to Testis)	• **Orchidectomy**

	• Orchiepididymitis • Orchioscirrhus • Orchitis
Organo- (related to Organ)	• Organofaction • Organogenesis • Organotherapy
Os(s)- (related to Bone)	• Ossicle • Ossification • Osteocyte • Osteoporosis • Osteoblast • Osteoclast • Osteology • Osteotomy
Osteo- (related to Bone)	• Osteoarthritis • Osteoporosis
Ot- (related to Ear)	• Otitis • Otoantritis • Otoscope
Ovario- (related to Ovary)	• Ovariectomy

(Table) cont.....

	• **Ovariotestis**
Ovo/i- (related to Egg)	• **Ovogenesis** • **Ovulation** • **Ovum** • **Ovigerm** • **Ovoplasm**
Palat- (related to Palate)	• **Palatography** • **Palatogram** • **Palatoplasty** • **Palatoscopy**
Pancreat(o)- (related to Pancreas)	• **Pancreatalgia Pancreatitis** • **Pancreatectomy**
Par(a)- (related to Beside, Near)	• **Parathyroid** • **Paraplegia** • **Paranasal**
Patho- (related to Disease)	• **Pathology** • **Pathogenesis** • **Pathophysiology**
Per- (related to Through, Throughout)	• **Percutaneous** • **Permeable**

	• **Perforation**
Phago- (related to Eating, Ingesting)	• **Phagocytosis** • **Phagocyte** • **Phagomania**
Pharyngo- (related to Pharynx, Throat)	• **Pharyngitis** • **Pharyngorrhagia** • **Pharyngoscope** • **Pharyngotomy**
Phlebo- (related to Vein)	• **Phlebectomy** • **Phlebitis** • **Phleborrhagia** • **Phlebotomy**
Phreno- (related to Diaphragm, Mind)	• **Phrenic** • **Phrenology** • **Schizophrenia** • **Phrenotomy**
Pilo- (related to Hair)	• **Pilocystic** • **Pilonidal** • **Piloerection** • **Pilosis**

(Table) cont.....

Pleuro- (related to Rib, Side, Pleura)	• **Pleuropneumonia** • **Pleuroscopy**
Pluri- (related to Many, More)	• **Pluripotent** • **Plurilateral**
Pneumono- (related to Lung)	• **Pneumonitis** • **Pneumonectomy**
Procto- (related to Anus, Rectum)	• **Proctologist** • **Proctoscopy**
Proto- (related to First, Primary)	• **Protoplasm** • **Protozoa** • **Protocell** • **Protolanguage**
Psycho- (related to Mind)	• **Psychics** • **Psychotherapy** • **Psychology**
Pyret- (related to Fever)	• **Pyrexia** • **Antipyretic**
Pyo- (related to Pus)	• **Pyocyte** • **Pyorrhea**
Rachio- (related to Spine)	• **Rachiocentesis**

(Table) cont.....

	• **Rachiodynia**
Recti/o- (related to Rectum)	• **Rectocele** • **Rectitis** • **Rectoscopy**
Rete/i- (related to Network)	• **Rete Testis** • **Retina** • **Reticulocyte**
Retro- (related to Backward, Behind)	• **Retrograde** • **Retrospective**
Rhin- (related to Nose)	• **Rhinoplasty** • **Rhinitis**
Sarco- (related to Flesh, Muscle)	• **Sarcoma** • **Sarcoplasm**
Sclero- (related to Hard, Hardening)	• **Sclerosis** • **Scleroderma**
Seps/t- (related to Infection)	• **Sepsis** • **Septicemia**
Sperm(at)o- (related to Sperm)	• **Spermatogenesis** • **Spermicide** • **Spermatocele**

	• **Spermagglutination**
Spiro- (related to Breathing)	• **Spirochete** • **Spirometry**
Spleno- (related to Spleen)	• **Splenomegaly** • **Splenectomy**
Stomato- (related to Mouth)	• **Stomatitis** • **Stomatology**
Supra- (related to Above, Over)	• **Supraorbital** • **Suprarenal**
Syn- (related to Together, With)	• **Syncytium** • **Synthesis** • **Syndrome** • **Synaptic** • **Synergy**
Thoraco- (related to Chest, Thorax)	• **Thoracotomy** • **Thoracic vertebrae**
Toxi/toxo- (related to Poison, Toxin)	• **Toxicology** • **Toxoplasmosis** • **Toxin**
Tracheo- (related to Trachea, Windpipe)	• **Tracheostomy**

	• **Tracheitis**
Tub- (related to Tube)	• **Tubectomy**
	• **Tubo-ovaritis**
	• **Tubulus**
Urethro- (related to Urethra)	• **Urethral**
	• **Urethritis**
	• **Urethrorrhagia**
	• **Urethrotomy**
Uro- (related to Urine, Urinary system)	• **Urobilin**
	• **Urology**
	• **Urogram**
	• **Uropenia**
Vaso- (related to Vessel)	• **Vasodilation**
	• **Vasospasm**
	• **Vasopressin**
Veno- (related to Vein)	• **Venous**
	• **Venography**
	• **Venopuncture**
	• **Venosclerosis**
Vesico- (related to Bladder)	• **Vesicoureteral**

(Table) cont.....

	• **Vesicotomy**
Vita- (related to Life)	• **Vitamin** • **Vital** • **Vitalism**
Zoo- (related to Animal)	• **Zoology** • **Zoologist** • **Zooparasite**
Zymo- (related to Enzyme)	• **Zymology** • **Zymogen** • **Zymoplasm**

BIBLIOGRAPHY

Alberts, B, Bray, D, Hopkin, K, Johnson, A.D Lewis, J, Raff, M, Roberts, K, Walter, P (2015) *Essential cell biology.* (2nd ed,). New York: Garland Science. 864.

Alley, H (2024) *The bee-keeper's handy book.*

Allison, LA (2021) *Fundamental molecular biology.* John Wiley & Sons.

Brachet, J & Mirsky, AE (1961) *The cell: biochemistry, physiology, morphology,* Volume III: meiosis and mitosis, Academic Press, pp. 77-412, ISBN: 978-0-12-123303-7.

Barillot, E, Calzone, L, Hupe, P, Vert, JP, Zinovyev, A (2013) *Computational systems biology of cancer.* Boca Raton, FL: CRC Press.

Bennett, J, Shostak, S, Schneider, N, MacGregor, M (2022) *Life in the universe.* Princeton University Press.

Bhatla, SC, Lal, MA (2023) *Plant physiology, development and metabolism.* Singapore: Springer.

[http://dx.doi.org/10.1007/978-981-99-5736-1]

Brevini, TA, Georgia, P (2012) *Gametogenesis, early embryo development and stem cell derivation, springer: briefs in stem cells* Springer Science & Business Media.

Bunyard, BA (2022) *The lives of fungi: a natural history of our planet's decomposers* 1st ed, Princeton University Press.

Butlin, R, Bridle, J, Schluter, D (2009) *Speciation and patterns of diversity..* Cambridge University Press.

Carlberg, C, Velleuer, E (2021) *Cancer biology: how science works.* Springer Cham.

http://dx.doi.org/10.1007/978-3-030-75699-4

Constable, PD, Hinchcliff, KW, Done, SH, Grünberg, W (2016) *Veterinary medicine: a textbook of the diseases of cattle, horses, sheep, pigs and goats.* Saunders Ltd.

Cornelissen, CN, Hobbs, MM (2012) *Lippincott illustrated reviews: microbiology (lippincott illustrated reviews series).* Lippincott Williams & Wilkins.

Cunningham, AJ (2012) *Understanding immunology.* Academic Press.

Durand, PM (2021) *The evolutionary origins of life and death.* University of Chicago Press.

Dyson, F (1999) *Origins of life.* Cambridge University Press.

http://dx.doi.org/10.1017/CBO9780511546303

Edelson, E (1999) Gregor Mendel: and the roots of genetics, Reprint edition. New York: Oxford University Press.

Fowler, S, Roush, R, Wise, J. (2024) *Concepts of biology.* OpenStax.

Gabriel, JA (2007) *The biology of cancer.* Wiley Online Library.

[http://dx.doi.org/10.1002/9780470988121]

Gahlawat, SK, Maan, S (2021) *Advances in animal disease diagnosis.* (1st ed., p. 336). Boca Raton: CRC Press.

[http://dx.doi.org/10.1201/9781003080282]

Gahlawat, SK, Maan, S (2021) *Advances in animal disease diagnosis.* (1ˢᵗ ed., p. 336). Boca Raton: CRC Press.

[http://dx.doi.org/10.1201/9781003080282]

Goodman, SR (2007) *Medical cell biology.* Academic Press.

Hierro, JL, Callaway, RM (2021) The ecological importance of allelopathy. *Annual Review of Ecology, Evolution, and Systematics, 52*(1), 25-45.

http://dx.doi.org/10.1146/annurev-ecolsys-051120-030619

Hudson, HJ (1986) *Fungal biology, CUP Archive.* (p. 298). London; Baltimore, Md, USA: Edward Arnold.

Hungerford, TG (1990) *Diseases of livestock* McGraw-Hill, Book Company (UK) Ltd.

Huston, MA (1994) *Biological diversity: the coexistence of species.* (p. 681). Cambridge; New York, NY, USA: Cambridge University Press.

Jain, HK, Kharkwal, MC (2004) *Plant breeding: mendelian to molecular approaches.* Springer Dordrecht.

[http://dx.doi.org/10.1007/978-94-007-1040-5]

Jorde, LB, Carey, JC, Bamshad, MJ (2019) *Medical genetics e-Book.* Elsevier Health Sciences.

Kar, D, Sarkar, S (2022) Genetics fundamentals notes 1st ed Springer, Singapore.

[http://dx.doi.org/10.1007/978-981-16-7041-1]

Kango, N (2013) *Textbook of microbiology.* IK International Pvt Ltd.

Lew, K (2018) *Taxonomy: the classification of biological organisms.* Enslow Publishing, LLC.

Lew, K & Jones, P (2021) Cell structure, processes, and reproduction, Chelsea House, pp. 127.

Liljas, A, Liljas, L, Lindblom, G, Nissen, P, Kjeldgaard, M, Ash, M (2016) *Textbook of structural biology.* World Scientific.

Malathi, V (1899) *Molecular biology.* Pearson Education India.

McMeekin, T, Mellefont, L, Ross, T (2007) *Predictive microbiology: past, present and future. Modeling Microorganisms in Food.* Woodhead Publishing.

Mire-Sluis, AR & Thorpe, R (1998) *Cytokines*, Academic Press, pp. 584, eBook ISBN: 9780080530215.

Money, NP (2014) *Microbiology: a very short introduction.* (p. 144). USA: Oxford University Press.

http://dx.doi.org/10.1093/actrade/9780199681686.001.0001

Mushtaq, W, Siddiqui, MB, Hakeem, KR (2020) *Allelopathy: Potential for green agriculture, Part of the book series: Springerbriefs in agriculture (briefsagro).* Springer Cham.

[http://dx.doi.org/10.1007/978-3-030-40807-7]

Ommati, MM, Retana-Márquez, R, Najibi, A, Heidari, R (2023) Advances in nanopharmacology: Focus on reproduction, endocrinology, developmental alterations, and next generational effects. *Nanopharmacology and nanotoxicology: Clinical implications and methods.* Bentham Science Publishers.

[http://dx.doi.org/10.2174/9789815079692123010008]

Ommati, MM., Heidari, R (2023) The role of taurine in the reproductive system: a focus on mitochondria-related mechanisms.*Taurine and the mitochondrion: applications in the pharmacotherapy of human diseases.* Bentham Science Publishers.

Ommati, MM., Heidari, R (2021) Amino acids ameliorate heavy metals-induced oxidative stress in male/female reproductive tissue.*Toxicology.* Academic Press.

[http://dx.doi.org/10.1016/B978-0-12-819092-0.00037-6]

Ommati, MM, Nozhat, Z, Sabouri, S, Kong, X, Retana-Márquez, S, Eftekhari, A, Ma, Y, Evazzadeh, F, Juárez-Rojas, L, Heidari, R, Wang, HW (2024) Pesticide-induced alterations in locomotor activity, anxiety, and depression-like behavior are mediated through oxidative stress-related autophagy: a persistent developmental study in mice. *Journal of Agricultural and Food Chemistry, 72*(19), 11205-11220.

http://dx.doi.org/10.1021/acs.jafc.4c02299] [PMID: 38708789]

Ommati, MM, Sabouri, S, Sun, Z, Zamiri, MJ, Retana-Marquez, S, Nategh Ahmadi, H, Zuo, Q, Eftekhari, A, Juárez-Rojas, L, Asefi, Y, Lei, L, Cui, S, Jadidi, MH, Wang, H, Heidari, R (2024) Inactivation of Mst/Nrf2/Keap1 signaling flexibly mitigates MAPK/NQO-HO1 activation in the reproductive axis of experimental fluorosis. *Ecotoxicology and Environmental Safety, 271*, 115947.

[http://dx.doi.org/10.1016/j.ecoenv.2024.115947] [PMID: 38215664]

Ommati, MM, Sabouri, S, Retana-Marquez, S, Nategh Ahmadi, H, Arjmand, A, Alidaee, S, Mazloomi, S, Akhlagh, A, Abdoli, N, Niknahad, H, Jamshidzadeh, A, Ma, Y, Azarpira, N, Asefi, Y, Heidari, R (2023) Taurine improves sperm mitochondrial indices, blunts oxidative stress parameters, and enhances steroidogenesis and kinematics of sperm in lead-exposed mice. *Reproductive Sciences, 30*(6), 1891-1910.

[http://dx.doi.org/10.1007/s43032-022-01140-5] [PMID: 36484981]

Ommati, MM, Ahmadi, HN, Sabouri, S, Retana-Marquez, S, Abdoli, N, Rashno, S, Niknahad, H, Jamshidzadeh, A, Mousavi, K, Rezaei, M, Akhlagh, A, Azarpira, N, Khodaei, F, Heidari, R (2022) Glycine protects the male reproductive system against lead toxicity *via* alleviating oxidative stress, preventing sperm mitochondrial impairment, improving kinematics of sperm, and blunting the downregulation of enzymes involved in the steroidogenesis. *Environmental Toxicology, 37*(12), 2990-3006.

[http://dx.doi.org/10.1002/tox.23654] [PMID: 36088639]

Ommati, MM, Arabnezhad, MR, Farshad, O, Jamshidzadeh, A, Niknahad, H, Retana-Marquez, S, Jia, Z, Nateghahmadi, MH, Mousavi, K, Arazi, A, Azmoon, MR, Azarpira, N, Heidari, R (2021) The role of mitochondrial impairment and oxidative stress in the pathogenesis of lithium-induced reproductive toxicity in male mice. *Frontiers in Veterinary Science, 8*(8), 603262.

[http://dx.doi.org/10.3389/fvets.2021.603262] [PMID: 33842567]

Ommati, MM, Li, H, Jamshidzadeh, A, Khoshghadam, F, Retana-Márquez, S, Lu, Y, Farshad, O, Nategh Ahmadi, MH, Gholami, A, Heidari, R (2022) The crucial role of oxidative stress in non-alcoholic fatty liver disease-induced male reproductive toxicity: the ameliorative effects of Iranian indigenous probiotics. *Naunyn-Schmiedeberg's Archives of Pharmacology, 395*(2), 247-265.

[http://dx.doi.org/10.1007/s00210-021-02177-0] [PMID: 34994824]

Ommati, MM, Shi, X, Li, H, Zamiri, MJ, Farshad, O, Jamshidzadeh, A, Heidari, R, Ghaffari, H, Zaker, L, Sabouri, S, Chen, Y (2020) The mechanisms of arsenic-induced ovotoxicity, ultrastructural alterations, and autophagic related paths: An enduring developmental study in folliculogenesis of mice. *Ecotoxicology and Environmental Safety, 204*, 110973.

[http://dx.doi.org/10.1016/j.ecoenv.2020.110973] [PMID: 32781346]

Pagano, M (2013) *Cell cycle control.* Springer Science & Business Media.

Paul, WE (2013) *Fundamental immunology.* Lippincott Williams & Wilkins.

Pecorino, L (2021) *Molecular biology of cancer: mechanisms, targets, and therapeutics.*

http://dx.doi.org/10.1093/hesc/9780198833024.001.0001

Pollard, TD, Earnshaw, WC, Lippincott-Schwartz, J, Johnson, G (2022) *Cell Biology E-Book.* Elsevier Health Sciences.

Quinn, PJ, Markey, BK, Leonard, FC, Hartigan, P, Fanning, S, Fitzpatrick, E (2011) *Veterinary microbiology and microbial disease.* Wiley-Blackwell.

Ragan, MA. (2023) *Kingdoms, Empires, and Domains: The history of high-level biological classification.* Oxford University Press.

[http://dx.doi.org/10.1093/oso/9780197643037.001.0001]

Ranta, E, Lundberg, P, Kaitala, V. (2005) *Ecology of populations.* Cambridge University Press.

[http://dx.doi.org/10.1017/CBO9780511610752]

Raoult, D, Fournier, PE, Drancourt, M (2004) What does the future hold for clinical microbiology? *Nature Reviews Microbiology, 2*(2), 151-159.

[http://dx.doi.org/10.1038/nrmicro820] [PMID: 15040262]

Rappole, JH (2022) *Bird migration: a new understanding.* Johns Hopkins University Press.

[http://dx.doi.org/10.1353/book.100160]

Rhodes, OE, Chesser, RK, Smith, MH (1996) *Population dynamics in ecological space and time.* University of Chicago Press.

Rizvi, S (2012) *Allelopathy: basic and applied aspects.* Springer Science & Business Media.

Rockwood, LL (2015) *Introduction to population ecology.* John Wiley & Sons.

Rossant, J, Tam, PT (2002) *Mouse development: patterning, morphogenesis, and organogenesis, San Diego, California.* London: Academic.

Ruddon, RW (2007) *Cancer biology.* Oxford University Press.

[http://dx.doi.org/10.1093/oso/9780195175448.001.0001]

Sapp, J (2009) *The new foundations of evolution: on the tree of life.* Oxford University Press.

Schatten, H, Schatten, G (1989) *The cell biology of fertilization.* Academic Press.

Schopf, JW (2002) *Life's origin: the beginnings of biological evolution.* University of California Press.

[http://dx.doi.org/10.1525/9780520928701]

Slack, JM., Dale, L (2021) *Essential developmental biology.* John Wiley & Sons.

Sinha, J, Bhattacharya, S (2006) *A Text book of Immunology.* Academic publishers.

Smith, RE (2014) *Medicinal chemistry-fusion of traditional and western medicine.* Bentham Science Publishers.

Sobel, JM, Chen, GF, Watt, LR, Schemske, DW (2010) The biology of speciation. *Evolution, 64*(2), 295-315.

[http://dx.doi.org/10.1111/j.1558-5646.2009.00877.x] [PMID: 19891628]

Spencer, H (2020) *The Principles of Biology.* Outlook Verlag.

Srinivas, S, Watanabe, T (2013) Early embryogenesis. *Textbook of Clinical Embryology.* (pp. 110-117). Cambridge: Cambridge University Press.

[http://dx.doi.org/10.1017/CBO9781139192736.014]

Thiagalingam, S (2015) *Systems biology of cancer.* Cambridge University Press.

[http://dx.doi.org/10.1017/CBO9780511979811]

Thorbjarnarson, J, Wang, X. (2010) *The Chinese alligator: ecology, behavior, conservation, and culture.* Johns Hopkins University Press.

[http://dx.doi.org/10.56021/9780801893483]

Tuchmann-Duplessis, H, Haegel, P (2013) *Organogenesis.* (Vol. II, p. 154). NY: Springer New York.

Vliet, KA (2020) *Alligators: the illustrated guide to their biology, behavior, and conservation.* Johns Hopkins University Press.

[http://dx.doi.org/10.1353/book.72661]

Walhout, M, Vidal, M, Dekker, J (2013) *Handbook of systems biology: concepts and insights.* Academic Press.

Webster, J, Weber, R (2007) *Introduction to fungi.* Cambridge University Press.

[http://dx.doi.org/10.1017/CBO9780511809026]

Weinberg, RA (2007) *The biology of cancer.* (1st ed., p. 864). New York: WW Norton & Company.

Wessner, D, Dupont, C, Charles, T, Neufeld, J (2020) *Microbiology.* John Wiley & Sons.

Winston, ML (1987) *The biology of the honey bee.* Harvard university press.

Woodger, JH (2014) Studies in the foundations of genetics. *Mathematics.* 27, 1959, 408-428.

Wusheng, J (2003) *English for students of veterinary science (Chinese).* (p. 247). Beijing, China: Agricultural Publishing House.

Wusheng, J (2010) *English for biology students (Chinese), 3rd ed* Higher Education Press, Beijing, China.

Yanagimachi, R (2017) Foreword. *The sperm cell: production, maturation, fertilization, regeneration.* Cambridge University Press.

[http://dx.doi.org/10.1017/9781316411124.001]

SUBJECT INDEX

A

Acid(s) 69, 70, 73, 74, 75, 78, 79, 80, 90, 96, 358
  deoxyribonucleic 70
  nucleic 69, 70, 73, 74, 75, 78, 79, 80, 90, 96, 358
Actin filaments 9
Adenosine 29
  diphosphate 29
  triphosphate 29
Agents, infectious disease 141
Air 27, 87, 109, 117, 134, 174, 204, 215
  contamination 134
  pollution, heavy 215
Algae, blue-green 10, 214
Anaerobic 91, 106
  heterotrophs 91
  respiration 106
Analysis of RNA transcripts 239, 376
Anaphase, mitotic 35
Animal(s) 156, 272
  embryos 156
  transgenic 272
Anthrax 136, 138
  transferred 136
Antibiotic 142, 272, 273
  development 272, 273
  resistance 142
Antigens, microbial 275

B

Bacteria 6, 29, 94, 98, 132, 133, 136, 137, 138, 139, 140, 214, 216
  photosynthetic 6
Bacterial endospores 134

C

Cancer 99, 222, 224, 225, 226, 227, 228, 229, 230, 231, 232, 234, 235, 237, 238
  biomarkers 235
  breast 227, 234, 235
  driver mutations 228, 229
  gene networks 228
  metastasis 228
  molecular networks 232
  prostate 227
  skin 99
Cellular 1, 5, 6, 8, 9, 10, 11, 28, 91, 106, 122, 143, 157, 237
  debris 10
  development 143
  functions 1, 6, 8, 9, 157, 237
  fungal filament 122
  homeostasis 8, 11
  membranes act 28
  motion 5
  respiration 91, 106
  signaling 10
Chargaff's 69, 71, 77, 82, 86, 357, 358
  observations 71, 86
  rules 69, 77, 82, 357, 358
Chemical reactions 7, 256, 260
Chemiosmosis 30
Chemotherapy 272, 273, 280
Chromosomal theory of inheritance 64, 355
Chromosome 38, 42, 43, 60
  pairs 42, 43
  segregation 38
  theory of heredity 60
Chytridiomycetes 116, 118
Clusters, ribosomal 17
Collagen degeneration 154
Conditions 25, 90, 91, 99, 103, 110, 111, 114, 115, 151, 226, 247, 275
  inflammatory 275
  regulated growth 151
Contemporary microbiology 140

Corticosteroids 36
Cosmologists 103
COVID-19 vaccines 276
Cyanobacteria 6
Cytokinesis 31, 33, 34, 35, 36, 37, 39, 40, 43,
    45, 46

D

Darwin's theory 195
Daughter cells 10, 32, 35, 36, 37, 38, 41, 57
    diverse 41
Daughter DNA duplex 87
Degradation 137, 142, 163, 240, 377
    pollutant 142
Dendritic cells 274, 275, 278, 279
    follicular 274
    immature 274
    mature 274, 275, 279
Digestion 8, 11, 18, 117, 124, 126, 137, 364
    bacterial 137
    cellular 8
    extracellular 117, 126
    intracellular 11
Digestive efficiency 272
Digital object identifier (DOI) 311, 312, 313,
    331, 332
Diseases 50, 133, 135, 136, 137, 138, 141,
    142, 200, 202, 222, 223, 247, 248, 249,
    250, 251, 270, 276
    autoimmune 141, 276
    cardiovascular 142
    -causing bacteria 138
    combat plant 141
    curbed 50
    foodborne 141
Disorders 38, 68, 87, 142
    autoimmune 142
    genetic 38, 68
    inherited 87
DNA 7, 13, 29, 37, 39, 43, 70, 71, 73, 74, 75,
    76, 77, 78, 86, 87, 91, 239, 240, 242,
    245, 271, 375, 376, 378
    and histone proteins 7
    and proteins 37, 239, 240, 242, 245, 375,
        376, 378
    damaged 242
    nucleus segregates 29
    of prokaryotic cells 13
    polymerase 74, 75

recombinant 271
replication and transcription 76
synthesis 74, 76
DNA and RNA 75, 112
    formation 112
    molecules 75
Double-helix configuration 70
Downloadable repository 290
Dutch elm disease 127
Dynamic 8, 226, 230, 231
    network visualization 231
    processes 8, 230
    systems theory 226

E

Electron microscopy 140
Electrophoresis 79, 358
Embryo 144, 145, 147, 150, 151, 152, 153,
    154, 155, 156, 157, 158, 159, 164, 166
    chicken 166
    developing 145, 151, 156, 158, 159
    growing 153
    mammalian 152
Embryonic development 146, 148, 151, 152,
    153, 154, 155, 156, 157, 162, 170, 174
Endoplasmic reticulum, non-ribosomal 16
Endosymbiotic Theory 4
Energy 7, 16, 29, 30, 73, 115, 130, 131, 132,
    139, 162
    harness 130
    metabolism 73
    process harnesses 30
Energy production 29, 141
    microbial 141
Environmental 117, 142, 192, 198, 219
    adaptations 192, 198, 219
    applications 142
    stress 117
Environments 6, 7, 19, 67, 68, 195, 197, 198,
    199, 201, 203, 205, 206, 212, 219, 220,
    228
    freshwater 19, 203, 205
    inflammatory 228
Enzyme(s) 8, 11, 13, 15, 69, 70, 75, 78, 81,
    87, 88, 91, 113, 239, 240, 241, 242, 243,
    245, 246, 348, 375, 376
    activity 69, 70
    deficiencies 87
    hydrolytic 8, 11, 13, 15, 348

-linked immunosorbent assay (ELISA) 239,
    240, 242, 245, 375, 376
rudimentary 113

F

Fermenting agents 125
Fertilization membrane 146, 368
Football coach tests 195
Framework, broader taxonomic 98
Function 6, 8, 70, 234, 358
  detoxification 8
  enzymatic 70, 358
  oncogenes 234
  sensory 6

G

Gametogenesis 144, 145, 163
Gametophytes 121
Gamtogenesis 35
Gases 111, 135, 153
  inorganic 111
  toxic 135
Gene(s) 37, 39, 141, 146, 157, 223, 224, 225,
    227, 230, 232, 233, 234, 235, 236, 237,
    239, 240, 241, 242, 243, 244, 375, 376,
    378
  activity 236, 237
  amplification 146, 157
  cancer driver-mutating 225, 237
  cancer-related mutated 234
  expression 37, 39, 223, 224, 232, 233, 239,
    240, 241, 242, 243, 244, 375, 376, 378
    function 141, 237
    microarray profiles 227, 230
  tumor-related 235
Genetic coding 74
Glycoprotein 125
Growth 32, 87, 116, 117, 134, 136, 152, 153,
    157, 170, 200, 201, 202, 203, 204, 205,
    223, 224, 227, 275
  bacterial 87
  cellular 32, 170
  embryonic 153
  lymphocyte 275
  tumor 227

H

Haploid cells 31, 36, 46, 68, 170
Health, cardiovascular 280
Heart transplants 50
Histone proteins 7, 45, 239, 241, 375, 376
Homeostasis 187
Hyperammonemia 313

I

Immune reactions 270
Immune responses 10, 226, 244, 270, 274,
    275, 278, 280
  cell-mediated adaptive 274
Infections 10, 129, 135, 136, 138, 142, 247,
    249, 276, 280
  bacterial 136
  fungal 135
  microbial 135
  systemic 247, 249
Infectious diseases 137, 141, 248, 270, 271
Influence 53, 68
  gene expression 68
  hereditary traits 53
Influenza 326, 327
  epidemics 326, 327
  virus 327
Information, metabolic 224
Inhibitors, immune checkpoint 276, 280
Inner cell mass (ICM) 147, 150
Inorganic nutrients 119, 128
Iron, oxidizing 139
Isolation 177, 179, 185, 270
  postzygotic 177, 179, 185
  techniques 270

J

Journal citation reports (JCR) 323, 326

K

Koch's postulates 136

L

Lipid(s) 3, 7, 20
  metabolizing 7

synthesis 3, 20
Liposomes 90, 98, 106
Lysosome packages 13

M

Meat 133, 134, 250
    decaying 133, 134
Mechanisms 1, 69, 72, 73, 76, 108, 113, 177,
        179, 190, 197, 199, 226, 227, 233
    bacterial DNA replication 73
    cellular mobility 1
    diverse 179
Mechanoenzymes 5
Memory cells 274, 278
Mendel's 52, 63
    laws 52
    principles 63
Metabolic 75, 77, 88, 91, 115, 248
    disease 248
    pathways 75, 77, 88, 91, 115
Meteoric bombardment 115
Microarray data analysis 234
Microbe(s) 131, 142, 270, 273
    infectious 273
    -microbe interactions 142
Microbial growth 134
Microbiology 141
    industrial 141
    medical 141
Microorganisms 129, 132, 133, 134, 135, 136,
        139, 140, 141, 142, 143
    isolated 136
    nurturing 133
Mitotic divisions 155, 170

N

Natural killer (NK) 273
Nitrogenous bases 70, 71, 73, 74, 75, 81, 358

O

Organic compounds 112, 131, 139
Organogenesis 144, 151, 152, 157

P

Paleontologists 180

Pathways, cancer signaling 226
Phosphoproteomic profiles 224
Photographic process 77
Physical 249, 271, 280
    fitness 249
    injuries 271
    therapy 280
Pinocytosis 3, 10, 12, 13, 346
Plants 121, 123, 202
    parasitic 121
    terrestrial 123, 202
    thallophyte 121
Process 70, 154, 157, 271, 276, 343
    biosynthetic 70
    developmental 154, 157, 343
    immune 271
    inflammatory 276
Protein(s) 1, 2, 3, 7, 8, 9, 10, 11, 12, 14, 18,
        29, 30, 74, 75, 79, 80, 99, 108, 114, 123,
        141, 238, 239, 240, 242, 243, 245, 276,
        346, 348, 358, 375, 376, 377, 378
    essential 123
    hemoglobin 79, 358
    inhibitors target 276
    membrane 30
    membrane-anchored 29
    membrane-embedded 30
    synthesis 1, 2, 3, 10, 11, 12, 14, 74, 141,
        346, 348
Proteinoid microspheres 104

R

Radiotherapy 273
Reproductive 93, 107, 178, 185, 198, 199
    isolation 93, 107, 178, 185, 198, 199
Resources 201, 202, 209, 211, 213, 258, 260,
        261, 262, 264, 266, 328, 344, 345, 373,
        374
    electronic 262
RNA 10, 229, 347
    interference (RNAi) 229
    ribosomal 10, 347

S

Science citation index (SCI) 252, 254, 265,
        266, 267, 312, 381
Signal transduction 239, 376
Signaling networks 229, 232, 233

Software 327
   licensing 327
   offerings 327
Soil 131, 139, 181, 296, 297
   bacteria 139
   environment 296, 297
   microbiology 139
Sperm mobility 163
Spermatocytes 145
Spermatogenesis 35, 144, 145, 158, 310, 311
Spermatogonia 35, 159, 366
Spermatozoa 35, 159

T

Techniques 133, 141, 275
   illumination 133
   molecular cloning 275
   recombinant DNA 141
Tooth decay 190, 191, 192
Transition, genetic 179
Tuberculosis 136, 137
Tumor microenvironment 228
Tumorigenesis 224

V

Vancouver method 306
Vesicles 10, 11
   double-membraned 11
   pinocytic 10
Virus, tobacco mosaic disease 137
Visual analysis 231
Visualization, chromosomal 70

W

Waste products 3, 11

X

X-ray 76, 77, 82, 273, 280, 357
   crystallography 273
   diffraction 76, 77, 82, 357
   imaging 280

www.ingramcontent.com/pod-product-compliance
Lightning Source LLC
Chambersburg PA
CBHW050758220326
41598CB00006B/53